汉英 锻压技术与装备词汇必备

洪慎章 胡卫诚 编

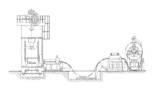

Chinese-English Dictionary of
Forging-Stamping
Technology and Equipment

上海交通大学出版社
SHANGHAI JIAO TONG UNIVERSITY PRESS

内容提要

本书收集了金属锻压常用专业词汇 5 000 余条,均选自英文锻压专业书籍、杂志及工程手册,包括工程基础、物理工程、金属塑性成形原理、锻件名称、冲压件名称、金属材料、钢锭冶炼、结构及缺陷、下料及检验、金属加热及加热炉、锻压设备及装置、锻造、冲压、锻压工艺、锻压模具、锻压辅助装置、锻压件缺陷、模具加工及机器、金属学及热处理、企业管理及组织,并附有 3 个附录:常用化学元素名称表,常用计量单位公制与英制换算表以及锻压设备技术参数举例。本书适合锻压行业专业人员、教师及学生使用。

图书在版编目(CIP)数据

汉英锻压技术与装备词汇必备/ 洪慎章,胡卫诚编.
—上海:上海交通大学出版社,2018
ISBN 978-7-313-20376-2

Ⅰ.①汉… Ⅱ.①洪… ②胡… Ⅲ.①锻压-词汇-汉、英②锻压设备-词汇-汉、英 Ⅳ.①TG31-61

中国版本图书馆 CIP 数据核字(2018)第 255834 号

汉英锻压技术与装备词汇必备

编　者:	洪慎章　胡卫诚		
出版发行:	上海交通大学出版社	地　址:	上海市番禺路 951 号
邮政编码:	200030	电　话:	021-64071208
出 版 人:	谈　毅		
印　制:	江苏凤凰数码印务有限公司	经　销:	全国新华书店
开　本:	850 mm×1168 mm　1/32	印　张:	6.25
字　数:	121 千字		
版　次:	2018 年 12 月第 1 版	印　次:	2018 年 12 月第 1 次印刷
书　号:	ISBN 978-7-313-20376-2/TG		
定　价:	38.00 元		

版权所有　侵权必究
告读者:如发现本书有印装质量问题请与印刷厂质量科联系
联系电话:025-83657309

前　言

《汉英锻压专业分类词汇》出版于1984年,时隔30多年,新技术不断更新,新名词日益增多。在此基础上,按锻压专业的各门课程及章节编写了《汉英锻压技术与装备词汇必备》一书。

本书共收集了常用专业词汇5 000余条,均选自英文锻压专业书籍、杂志及工程手册。本书共分19章,内容包括：工程基础、物理工程、金属塑性成形原理、锻件名称、冲压件名称、金属材料、钢锭冶炼、结构及缺陷、下料及检验、金属加热及加热炉、锻压设备及装置、锻造、冲压、锻压工艺、锻压模具、锻压辅助装置、锻压件缺陷、模具加工及机器、金属学及热处理、企业管理及组织。附编了3个附录：① 常用化学元素名称表；② 常用计量单位公制与英制换算表；③ 锻压设备主要技术参数举例。

全书是由上海交通大学材料科学与工程学院塑性成形技术与装备研究院洪慎章教授策划编写,上海实业交通电器有限公司胡卫诚工程师修改及补充。由上海交大中京锻压有限公司给于出版赞助,在此表示感谢。

由于编者的水平有限,书中不妥和错误之处在所难免,恳请广大读者不吝赐教,以便得以修正,以臻完善。

编　者

2018年8月

目 录

- 01 工程基础 ··· 001
- 02 物理工程 ··· 009
- 03 金属塑性成形原理 ······································ 019
  - 3.1 术语 ··· 019
  - 3.2 变形力学 ··· 020
  - 3.3 力学性能与塑性变形机理 ···················· 023
- 04 锻件名称 ··· 026
  - 4.1 汽车,拖拉机,机床 ······························ 026
  - 4.2 飞机,船舶,武器 ································· 036
  - 4.3 五金工具,日用器皿 ···························· 040
- 05 冲压件名称 ·· 044
  - 5.1 汽车,拖拉机,摩托车,自行车 ··············· 044
  - 5.2 飞机,船舶,武器 ································· 045
  - 5.3 家用电器,厨房电器 ···························· 045
  - 5.4 五金工具,日用器皿 ···························· 047
- 06 金属材料 ··· 049
- 07 钢锭冶炼、结构及缺陷 ······························ 060

08	下料及检验 ……………………………………	070
09	金属加热及加热炉 ……………………………	078
	9.1　加热 ……………………………………	078
	9.2　加热炉 …………………………………	080
10	锻压设备及装置 ………………………………	089
	10.1　锻造设备及零件 ………………………	089
	10.2　冲压设备及零件 ………………………	099
11	锻造 ……………………………………………	105
	11.1　自由锻 …………………………………	105
	11.2　模锻 ……………………………………	110
12	冲压 ……………………………………………	112
13	锻压工艺 ………………………………………	115
	13.1　模锻工艺 ………………………………	115
	13.2　冲压工艺 ………………………………	119
14	锻压模具 ………………………………………	124
	14.1　锻造模具 ………………………………	124
	14.2　冲压模具 ………………………………	126
	14.3　其他模具 ………………………………	129
15	锻压辅助装置 …………………………………	132
	15.1　摩擦与润滑 ……………………………	132
	15.2　清理装置 ………………………………	133
	15.3　机械化与自动化 ………………………	134
16	锻压件缺陷 ……………………………………	137

  16.1 锻件缺陷 ·················· 137

  16.2 冲压件缺陷 ················ 139

17 模具加工及机器 ················ 141

18 金属学及热处理 ················ 149

19 企业管理及组织 ················ 157

附录 ························ 171

附录1 常用化学元素名称表 ··········· 173

附录2 常用计量单位公制与英制换算表 ······ 176

附录3 锻压设备技术参数举例 ··········· 178

参考文献 ······················ 189

01 工程基础　Engineering Basis

百分数,百分率	percentage
半径	radius
比例	proportion
标准,规格	standard
表	list
价目表	price list
零件表	parts list
不等式	inequality
草图,略图	sketch
长度	length
长方形	rectangle
垂直,正交	perpendicular
次序,日程,等级,规则	order
粗糙度	roughness
代数	algebra
度,次	degree
对称	symmetry
对数	logarithm

多边形	polygon
方程式	equation
方向	direction
分类	classification
分母	denominator
分析	analyse/analyze
分子	numerator
符号,标记	mark/symbol/sign
符合	accord with
复数	plurality
概率	probability
概念	concept
高度	height
公差	tolerance
公式	formula
估算	estimate
光洁度	fineness
光泽度,光滑度	gloss
过渡	transition
过盈	interference
函数	function
行列式	determinant
厚度	thickness
弧度	radian

极限	limit
几何学	geometry
计算	calculation
反复计算	repeated calculation
简化计算	short-cut calculation
矩阵计算	matrix calculation
普通计算,常规计算	routine calculation
图解计算	graphic calculation
间隙	gap
检查	check
检修	examine and repair
检验	test
减轻	lessen
减少,减小	reduce
简图,方案	scheme
角,角度	angle
弧角	arc angle
圆角	rounded angle
精确度	accuracy/precision
径向	radial direction
距离	distance
刻度,比例尺	scale
华氏温度表	Fahrenheit thermometric scale
开氏温度表,绝对温	Kelvin thermometric scale

度表	
摄氏温度表	Celsius thermometric scale
宽度	breadth/width
蓝图	blueprint
棱柱	prism
立体	solid
逻辑	logic
密度	density
面	face
侧面	lateral face
承压面	pressure face
顶点面	vertex face
端面	end face
工作面	working face
接触面	contact face
接合面	joint face
控制面	control face
内表面	inner face
前面	front face
外表面	outer face
锥面	pyramidal face
面积	area
命名,名称	nomination
命名法	nomenclature

模型, 样机	model
排列	permutation
配合	coordinate
平面	plane
平行的, 并联的, 同样的	parallel
切向	tangent direction
球体	sphere
区域, 范围	region
全套, 一组	set/suit
三角形	triangle
三角学	trigonometry
深度	depth
升高, 提高	raise
式, 表达式	expression
视图	view
侧视图	lateral view
底视图	bottom view
断面图	sectional view
放大图	enlarge view
俯视图	vertical view
后视图	back view
局部放大图	partical enlarge view
平面图	plan view
剖视图	cut-open view

全视图	full/general view
仰视图	upward view
右视图	right view
左视图	left view
数量	quantity
说明书,指南	instruction
四边形	quadrilateral
特征,性质	character
提纲	outline
体积,容量,容积	volume
图	figure
图表	diagram
外形	configuration
温度计	thermometer
物质,本质	substance
系数	coefficient
线	line
尺寸线	dimension line
地平线	horizon line
对角线	diagonal line
弧线	arc line
基线	base line
平衡线	balance line
切线	tangent line

实线	actual/real line
水平线	horizontal line
同轴线	coaxial line
斜线	oblique line
虚线	imaginary line
直线	straight line
指示线	index line
中线,中间线	median line
中心线	center line
形式,方式	mode
形状,形态	shape
性能,状态	behavior
循环	circulation
验证	verify
因素	factor
余割	cosecant
余切	cotangent
余弦	cosine
原始的,最初的	original
圆,圆周	circle
圆筒	cylinder
圆柱体	cylinder
圆锥	circular cone
圆锥体	cone

增加,增长	increase/add/grow
正方形	square
正割	secant
正切	tangent
正弦	sine
直径	diameter
指数	exponent/index
质量	quality
种,等级,类目	class
轴,轴线	axis
横轴	horizontal axis
纵轴	direct axis
轴向	axial direction
组合	combination
最佳的,最优的,最适的	optimum
坐标	coordinate
横坐标	horizontal coordinate
直角坐标	rectangular coordinate
纵坐标	longitudinal/vertical coordinate

02 物理工程 Physical Engineering

摆动,回转	swing
保存,守恒	conservation
比热	specific heat
比容	specific capacity/specific volume
比重	specific weight
变化	change
变压器	transformer
波	wave
磁波	magnetic wave
电波	electric wave
电子波	electron wave
光波	light wave
气波	aerodynamic wave
声波	acoustic wave
玻璃	glass
泊松比	Poisson's ratio
补偿	compensation
补充	replenish

参数	parameter
差,差别,区别	difference
差异,对比	contrast
缠绕	twine
充填	filling
抽出	pumping out
传播,扩散	propagation
传送,传导,转换,变换	transfer
吹,打击	blow
磁场	magnetic field
磁针	magnetic needle
催化作用,触媒作用	catalysis
大气	atmosphere
等密度线	isodense
等容线	isoster
等温线	isotherm
等压线	isobar
等值线	isopleth
电池	battery
电动机	motor
电极	electrode
电解	electrolysis
电流	electric current
电压	voltage

电源	power supply
电子	electron
电阻	resistance
定律,法则	law
胡克定律	Hooke's law
牛顿万有引力定律	Newton's law of universal gravitation
动力学	kinetics
动量	momentum
动作,作用	action
断面收缩率	shrinkage on cross section
对流,传递	convection
发电机	generator
发动机	engine
反应	reaction
非稳定性	non-stability
沸腾	boiling
分解	decomposition
分子	molecule
分子的	molecular
分子力	molecular force
分子量	molecular weight
分子式	molecular formula
粉末	powder
浮力	buoyancy

符合,一致,对应	correspondence
辐射	radiation
干扰	disturbance/interference
高分子	high molecular
供应,调节	accommodation
共振	resonance
固态	solid state
固体	solid body
光谱	spectrum
光学	optics
轨迹,路线	path
合成	synthesis
化合,组合	combination
化合物	compound
回响	reverberation
混合	compound
混合物	mixture
火花	spark
积累,储存	accumulation
激光	laser
挤压	extrusion
加速	acceleration
减速	deceleration
交叉	crossover

交流	alternate current
搅拌	agitation
阶段,程度	stage
酒精	alcohol
绝对	absolution
绝对零度	absolute zero
绝热的	adiabatic
空间	space
蜡,石蜡	wax
力	force
反作用力	counter force
惯性力	inertial force
离心力	centrifugal force
平衡力	balance force
向心力	centripetal force
引力	attraction force
质量力	mass force
作用力	applied force
力学	mechanics
流明	lumen
媒介	medium
煤油	kerosene
磨损	wear
木材	wood

能，能量	energy
能力	ability
黏着，黏附	adhere
扭矩	torque
传动扭矩	driving torque
旋转扭矩	rotary torque
制动扭矩	brake torque
抛	throw
喷射	spray
疲劳极限	fatigue limit
偏差	deviation
频率	frequency
气态	gaseous state
气体	gas
汽油	benzine
强度	strength
冲击强度	impact strength
规定强度	proof strength
极限强度	ultimate strength
均匀强度	uniform strength
抗拉强度	tensile strength
抗扭强度	torsional strength
抗弯强度	bending strength
抗压强度	compression strength

疲劳强度	fatigue strength
屈服强度	yield strength
蠕变强度	creep strength
取，拿	take
缺口	notch
扰动	disturbance
韧度	toughness
扫描	scan
伸长率	elongation percentage
声谱	sound spectrum
声音	sound
失效，故障	failure
矢量，向量	vector
试样，试件	specimen
树脂	resin
刷	brush
水泥	cement
速度，比例，比率	rate
塑料	plastics
热固性塑料	thermosetting plastics
热塑性塑料	thermoplastic plastics
缩颈	necking
陶瓷	ceramics
天线	antenna

添加,附加	addition
同步	synchronization
投影	projection
涂,擦,敷	smear
稳定性	stability
吸收	absorption
吸引	attraction
显示,指示	display
线圈	coil
并联线圈	parallel coil
串联线圈	series coil
相对	relative
效率,功率	efficiency
效应,作用,影响	effect
性质	quality
压制	pressing
阳极	positive pole
咬住,夹紧	bite
液态	liquid state
液体	liquid
阴极	negative pole
音调	tone
油,润滑油,石油	oil
柴油	diesel oil

淬火油	quenching oil
废油	waste/used/refuse/slop oil
航空油	aeroplane oil
棉籽油	cottonseed oil
汽缸油	cylinder oil
燃油	burning oil
原油	crude oil
重油	black oil
轴承油	bearing oil
原子	atom
原子核	atomic nucleus
运动	motion
分子运动	molecular motion
交叉运动	crisscross motion
均速运动	uniform motion
连续运动	continuous motion
螺旋运动	helical motion
往复运动	alternating motion
谐和运动	harmonic motion
圆周运动	circular motion
直线运动	rectilinear motion
轴向运动	axial motion
组合运动	aggregate motion
运动学	kinematics

噪音	noise
真空	vacuum
振动	vibration
振幅	amplitude
蒸发	vapor
直流	direct current
质子	proton
周期	period
注射	inject
转换,变成	conversion
状态,情况	state
非稳定状态	unsteady state
过渡状态	transition state
临界状态	critical state
稳定状态	steady state
亚稳定状态	metastable state

03 金属塑性成形原理 Principles of Metal Plastic Forming

3.1 术语　　Term

板料成形	sheet forming
初次成形加工	primary metalworking
等温锻	isothermal forging
等温模锻	isothermal die forging
锻压	forging and stamping
二次成形加工	secondary metalworking
金属回转加工	rotary metalworking
金属塑性加工	metal plastic working
金属压力加工	metal pressure working
冷锻	cold forging
热锻	hot forging
体积成形	bulk forming
温锻	warm forging
无屑加工	chipless working

3.2 变形力学   Deformation Mechanics

变形	deformation
变形功	deformation work
变形力	deformation force
变形力计算的变分法	variational method of solution of the deformation load/force
变形力计算的工程计算法	engineering method of solution of the deformation load/force
变形力计算的上限法	upper-bound method of solution of the deformation load/force
变形力计算的下限法	lower-bound method of solution of the deformation load/force
变形力计算的主应力法	slab method of solution of the deformation load/force
不均匀变形	non-homogeneous deformation
单位变形力	unit deformation pressure
弹塑性有限元法	elastic-plastic finite element method
动可容速度场	kinematically admissible velocity field
非稳定变形过程	non-steady deformation process
刚塑性有限元法	rigid-plastic finite element method
高斯应力曲面	Gauss stress quadric
光塑性法	photoplasticity method
亨盖第二定理	Hencky's second theorem
亨盖第一定理	Hencky's first theorem

亨盖积分	Hencky's integration
滑移线场理论	slip line field theory
金属变形	metal deformation
静可容应力场	statically admissible stress field
均匀变形	homogeneous deformation
柯西问题	Cauchy problem
黎曼问题	Riemann problem
理想刚塑性体	rigid-perfectly plastic body
理想弹塑性体	elastic-perfectly plastic body
米塞斯圆柱面	Mises cylinder
密栅云纹法（莫尔云纹法）	Moire method
能量准则	energy criterion
平均压力	mean pressure
平面问题	plane problem
平面应变状态	plane strain state
平面应力状态	plane stress state
球应力张量	spherical stress tensor
屈服曲面	yield surface
屈服准则	yield criteria
屈雷斯加六棱柱面	Tresca hexagonal prism
全量理论	total strain theory
三向应力状态	three-dimensional stress state
伸长类变形	tensile type of deformation

死区	dead metal
速度间断	velocity discontinuity
速端图	hodograph
塑性铰	plastic hinge
塑性势	plastic potential
特征线法	characteristic line method
体积不变条件	constancy of volume
稳定变形过程	steady deformation process
压缩类变形	compressive type of deformation
应变	strain
等效应变	equivalent strain
对数应变	logarithmic strain
工程应变	engineering strain
名义应变	nominal strain
真实应变	true strain
应变速率	strain rate
应变状态	strain state
应力	stress
八面体剪应力	octahedral shear stress
八面体正应力	octahedral normal stress
残余应力	residual stress
等效应力	equivalent stress
附加应力	additional stress
静液应力	hydrostatic stress

名义应力	nominal stress
真实应力	true stress
主应力	principal stress
最大剪应力	maximum shear stress
应力莫尔圆	Mohr's circle of stress
应力偏量	stress deviation
应力偏量不变量	invariant of stress deviation
应力应变曲线	stress-strain curve
应力张量	stress tensor
应力张量不变量	invariant of stress tensor
应力状态	stress state
永久变形	permanent deformation
有心扇形场	centred-fan field
增量理论	incremental strain theory
直观塑性法	visioplasticity method
轴对称问题	axisymmetrical problem
最大剪应力准则	maximum shear stress criterion
最小阻力定律	law of minimum resistance
坐标网格法	coordinate grid method

3.3 力学性能与塑性变形机理 Mechanical Property and Plastic Deformation Mechanism

| 包辛格效应 | Bauschinger effect |
| 超塑性 | superplasticity |

迟滞性	hysteresis
断裂韧性	fracture toughness
锻造流线	forging flow line
二次再结晶	secondary recrystallization
高速脆性	high velocity brittle
各向同性	isotropy
各向异性	anisotropy
厚向异性指数	the strain ratio of width and thickness
滑移变形	slide deformation
滑移线	slip line
回复	recovery
加工硬化指数	work-hardening exponent
抗氧化性	oxidation resistance
扩散蠕变	diffusion creep
蓝脆性	blue brittleness
冷变形强化	cold deformation strengthening
冷脆性	cold brittleness
力学性能	mechanical property
临界变形程度	critical deformation
螺旋位错	helical/sprical dislocation
名义屈服极限	nominal yield stress
屈服平台	yield point elongation
屈强比	yield to tensile ratio

热脆性	hot brittleness
热强性	heat resistance
刃型位错	blade dislocation
塑性	plasticity
塑性图	diagram of forgebility
位错攀移	dislocation climb
形变织构	deformation texture
应变速率敏感性指数	strain-rate sensitivity exponent
应变速率强化效应	effect of strain rate strengthening
应力强度因子	stress intensity factor
应力松弛	relaxation of stress
硬化曲线	curve of hardening
再结晶	recrystallization
再结晶温度	recrystallization temperature

04 锻件名称 Forging Designation

4.1 汽车,拖拉机,机床　　Automobile, Tractor, Machine

臂,杆,柄	arm
操纵杆	control arm
叉形臂	fork arm
导杆	guide arm
调速器杆	governor arm
横杆,横臂,悬臂	cross arm
离合器杆	clutch arm
连杆臂	link arm
履带轮臂	track wheel arm
气阀摇臂	valve rocker arm
前保险杆	front bumper arm
曲柄臂,连杆	crank arm
踏板臂	pedal arm
羊角臂,转向节臂	knuckle arm
摇臂	rocker arm
支架,悬臂	ally arm

制动臂	brake arm
制动摇臂	brake rocker arm
主转向臂,主操纵臂	main steering arm
转动臂	rotor arm
转向臂	steering arm
转向垂臂	pitman arm
转向摇臂	steering knuckle arm
操纵杆	controller
变速操纵杆	variable speed controller
叉	fork
半轴叉	half axle fork
变速叉	transmission fork
车架叉	frame fork
齿轮换挡叉	gear shift fork
倒车换挡叉	reverse gear shift fork
调速器叉	governor fork
分离叉	disengaging fork
分离叉	release fork
高速换挡叉	high speed shift fork
焊接叉	weld fork
滑动叉	sliding fork
接合(器)叉	coupling fork
离合(器)叉	clutch fork
离合器分离叉	clutch release fork

螺纹叉	thread fork
前轴叉	front axle fork
三挡及四挡换挡叉	third and forth speed shift fork
头挡及二挡换挡叉	first and second speed shift fork
凸缘叉	flange fork
万向节叉	universal joint fork
衬套	bush
齿轮轴套	gear shaft bush
隔套	distance bush
花键轴套	castellated shaft bush
换向器套筒	commutator bush
齿轮	gear
变速齿轮	speed change/variable gear
变速轮,变向轮	change gear
并联齿轮,组合齿轮	combination gear
差动齿轮	differential gear
从动齿轮	driven gear
从动螺旋齿轮	driven helical gear
从动伞齿轮	driven cone gear
倒车齿轮	reverse gear
动力齿轮	power gear
惰轮,空转轮	idle gear
固定齿轮	fixed gear
滑动齿轮	sliding gear

螺旋齿轮	helical gear
内齿轮	annular gear
啮合齿轮	engaging gear
驱动齿轮	driving gear
人字齿轮	chevron/double helical/herringbone gear
伞齿轮,圆锥齿轮	bevel/cone gear
塔齿轮	stepped gear
外齿轮	external gear
蜗轮	worm gear
行星齿轮	planet gear
正齿轮	spur gear
直齿轮	straight gear
直齿伞齿轮	straight bevel gear
主动齿轮	drive gear
主动螺旋齿轮	drive helical gear
主动伞齿轮	drive cone gear
盖,帽	cap
连杆盖	connecting rod cap
杆,杠杆,柄	lever
调节杆	adjustable/regulating lever
分离杆	release lever
离合器杆	clutch lever
离合器脚踏杆	clutch foot lever

启动杆	actuating lever
手柄	hand lever
制动手柄	brake lever
转向臂	steering lever

杆 bar
操作杆	action bar
杠杆	pry bar
缓冲杆	buffer bar
拉杆	pull/draft bar
连接杆	coupling bar
推杆	push bar
拖杆	tow bar
止推杆	thrust bar
制动杆	brake bar

杆 rod
连杆	connecting rod
推杆	push rod
转向拉杆	steering drag rod

钩 hook
| 前拖钩 | front tow hook |
| 拖车钩 | car puller hook |

活门,阀 valve
| 进气阀 | air inlet valve |
| 排气阀 | exhaust valve |

夹盘，卡盘	chuck
接头	joint
法兰接头	flange joint
万向接头	universal joint
连杆，杆，环节	link
传动杆	drive link
惰性连杆	lazy/inertness link
拉杆	drag/draw link
伸缩杆	expansion link
十字头连杆	crosshead link
万向节轴	cardan link
摇臂连杆	rocker arm link
主连杆	master link
转向拉杆	steering drag link
转向连杆	steering link
联轴节	coupling
联轴套	coupling band
轮毂	hub
飞轮毂	flywheel hub
凸缘毂	flange hub
十字轴	cross shaft
踏板	pedal
离合器踏板	clutch pedal
凸轮，偏心轴	cam

导向凸轮	guide cam
进气凸轮	admission cam
排气凸轮	exhaust cam
偏心凸轮	eccentric cam
心形凸轮	frog cam
制动凸轮	brake cam
凸轮轴	camshaft
凸缘,法兰	flange
法兰模板,凸缘体	body flange
固定凸缘	holding flange
管凸缘	pipe flange
接合凸缘	matching flange
接头凸缘	joint flange
肋凸缘	rib flange
套筒法兰	sleeve flange
凸缘法兰	male flange
轴法兰	shaft flange
万向接头	cardan joint
圆盘	disk
阀盘	valve disk
分离盘	seperating disk
号码盘	number disk
检验盘	monitor disk
卡盘	clamping disk

刻度盘	index disk
喇叭盘	horn disk
离合器盘	clutch disk
轮盘	wheel disk
起动圆盘	actuator disk
扇形盘	sector disk
凸轮盘	cam disk
制动盘	brake disk
轴法兰盘	shaft disk
爪形盘	claw disk
转盘	swivel disk
轴	axis
万向节轴	cardan axis
主轴	principal axis
转轴	spin axis
轴,车轴	axle
摆动轴	pivoted axle
半轴	half axle
不连轴,自由轴	uncoupled axle
叉形轴	forked axle
传动轴,传力轴	live axle
端轴	end axle
鼓轮轴	drum axle
横轴	lateral axle

后车桥，后桥	back axle
后轴	rear axle
铰接后轴，铰接后桥	articulated rear axle
空心轴	hollow axle
连动轴	couple axle
链轴	chain axle
前梁，前轴	front axle
软轴	flexible axle
套筒轴	telescopic axle
万向轴，铰接轴	cardan axle
旋转轴	turning axle
移动轴，换向轴	shifting axle
支承轴	supporting axle
中间轴	intermediate axle
主动轴	leading axle
转动轴	rotating axle
转向摇臂轴	steering rocker arm axle

轴	shaft
变速叉轴	gear shift fork shaft
变速杆轴	gear shift lever shaft
变速器中间轴	transmission counter shaft
车轴	axle shaft
齿轮轴	gear shaft
从动轴	driven shaft

电枢轴	armature shaft
调速器轴	governor shaft
动力轴,传力轴	power shaft
分配轴	distributing shaft
副轴	auxiliary shaft
杠杆轴	lever shaft
固定轴	stationary shaft
后传动轴	inter-axle shaft
后轮轴,后桥轴	rear axle shaft
花键轴	castellated shaft
花键轴	spline shaft
连接杆轴	pitman shaft
连接轴	connection shaft
联结轴	coupling shaft
偏心轴	eccentric shaft
台阶轴	stepped shaft
凸轮轴	cam shaft
万向传动轴	universal driving shaft
万向联接轴	universal jointed shaft
万向轴	cardan shaft
蜗杆轴	worm shaft
压缩机轴	compressor shaft
制动横轴	brake cross shaft
制动凸轮轴	brake cam shaft

制动中轴	brake intermediate shaft
中空轴,套筒轴	quill shaft
主传动轴	main drive shaft
主动轴	drive shaft
转向轴	steering shaft
组合轴	built-up shaft
轴,心轴	spindle
花键轴	spline spindle
镗杆	boring spindle
套轴	sleeve spindle
凸轮轴	cam spindle
凸缘轴	flange spindle
铣床副轴	auxiliary milling spindle
轴颈	axle spindle
钻轴	drill spindle
轴承	bearing
肘节,钩爪	knuckle
转向节	steering knuckle

4.2 飞机,船舶,武器   Aircraft, Shipping, Weapon

车架	car frame
大梁	large beam
带法兰阀体	valve body of flange

导弹	missile
阀盖	valve cap
阀体	valve body
飞机螺旋桨	airscrew
风扇	fan
风扇轮毂	fan hub
封头	head plate
隔板,舱壁	bulkhead
横梁	cross beam
滑轨,履带	track
环,圈,轮	ring
保护环	guard ring
炮尾环	breech ring
火箭	rocket
机架	frame
卡板	baffle
连接板	splice plate
链轮	sprocket
轮	wheel
带叶片的整体涡轮	integral turbine wheel with blades
轮轴	wheel axle
起落架轮	landing wheel
叶轮	blade wheel
螺旋桨,推进器	propeller

螺旋桨轴	propeller shaft
密封盖,封头	sealed cover
面板,翼片	panel
内侧板	inner panel
炮管	gun barrel
起落架	landing gear
起重机吊钩	crane hook
前梁	front beam
潜水艇	submarine
枪,炮	gun
曲柄箱	crankcase
曲轴	crankshaft
单拐曲轴	one-throw crankshaft
双拐曲轴	double-throw crankshaft
整体曲柄轴	integral crankshaft
组合式曲轴	built-up crankshaft
双角吊钩,山形钩	double hook
筒,柄	stem
外侧板	outer panel
尾轴	tail shaft
涡轮盘	turbine disk
涡轮轴	turbine shaft
压气机盘	compressor disk
叶轮	impeller

叶片	blade
导向叶片	guide blade
定子叶片	stater blade
风扇叶片	fan blade
固定叶片	fixed blade
螺旋桨叶片	propeller blade
扭转叶片	twisted blade
平叶片	flat blade
透平叶片	turbine blade
凸出叶片	bulged blade
压气机叶片	compressor blade
叶背	blade back
叶根	blade root
叶尖	blade tip
叶轮叶片	impeller blade
叶盆	blade basin
翼梁	spar (beam)
后梁	rear spar (beam)
假梁	false spar (beam)
前梁	front spar (beam)
鱼雷气瓶	torpedo gas bottle
轧辊	roll
支承	support
支架,支座	bracket

支柱	pillar
中间轴	intermediate shaft
轴颈	shaft journal
主梁	main beam
贮器	reservoir
转子	rotor
桨毂	rotor hub

4.3 五金工具，日用器皿　　Hardware Tools, Daily Utensils

扳手	spanner/wrench
叉形扳手	fork spanner/wrench
单头呆扳手	single end spanner/wrench
法兰扳手	flange spanner/wrench
方头扳手	square spanner/wrench
钩形扳手	hook spanner/wrench
管子扳手	tube spanner/wrench
活扳手	adjustable spanner/wrench
棘轮扳手	ratchet spanner/wrench
两用扳手	combination spanner/wrench
螺帽扳手	nut spanner/wrench
螺丝扳手	screw spanner/wrench
内六角扳手	inner hexagon ring spanner/wrench
双头呆扳手	double end spanner/wrench

双头梅花扳手	double end hexagon ring spanner/wrench
套筒扳手	socket spanner/wrench
歪柄扳手	skew spanner/wrench
弯头扳手	bent spanner/wrench
直柄扳手	straight spanner/wrench
扁斧,横斧	adze
刨刀	planer tool
锄头	hoe
锤	hammer
八角锤	blacksmith hammer
钳工锤	machinist hammer
石工锤	stone hammer
羊角锤	claw hammer
圆头锤	ball hammer
锉	file
半圆锉	half-round file
扁锉	equaling/flat file
方锉	square file
尖锉	entering file
三角锉	triangle file
什锦锉	broach file
圆锉	round file
雕刻刀	burin

斧	axe
锤斧	hammer axe
剁斧	chip axe
木工斧	broad axe
手斧	hand axe
钢叉	steel fork
镐	pick
风镐	air/miners pick
夯镐	tamping pick
煤镐	coal pick
平头镐	flat pick
土镐	clay pick
击剑	fencing
剪刀	scissors
剑,刀	sword
犁刀	coulter
镰刀	sickle
轮圈刹车	rim brake
木工车刀	wood turning tool
木钻	wood drill
钳,夹钳	pliers
扁嘴钳	flat nose pliers
电工钳	electrician pliers
钢丝钳	cutting (wire) pliers

管子钳	pipe pliers
胡桃钳	walnut pliers
尖嘴钳	long nose pliers
开箱钳	unpack pliers
鲤鱼钳	slip joint pliers
手钳	hand pliers
水泵钳	water pump pliers
台钳	bench pliers
圆嘴钳	round nose pliers
手术刀	scalpel
手闸	handbrake
錾刀	engraving tool
凿	chisel
扁尖凿	bolt chisel
木凿	wood chisel
平凿	flat chisel
钳工凿	bench chisel
手凿	hand chisel
中心凿	center chisel
爪	pawl
回转爪	rotary pawl
夹爪	grip pawl
绞车爪	winch pawl
制动爪	locking pawl

05 冲压件名称 Stamping Designation

5.1 汽车,拖拉机,摩托车,自行车　Automobile, Tractor, Motorcycle, Bicycle

把手	handle
车顶	car roof
车身	car body
挡泥板	mudguard
灯罩	lamp cover
钢圈	steel ring
后盖	back cover
立管	stand pipe
轮毂	hub
内饰件	interior decoration parts
前保险杠	front bumper
前盖	front cover
外门板	back head
外饰件	exterior decoration parts
仪表盘	dashboard
油箱	fuel tank

| 座椅架 | seat frame |

5.2 飞机,船舶,武器　　Aircraft, Shipping, Weapon

步枪	rifle
船身	ship body/hull
弹壳	shell
飞机骨架	airframe
机关枪	machine gun
机翼	airfoil
卡宾枪	carbine
蒙皮	skin
手枪	pistol
圆筒	cylinder
缸体,筒体	cylinder body

5.3 家用电器,厨房电器　　Household Electric Appliances, Kitchen Electric Appliances

冰箱	refrigerator
冷藏室	refrigerating chamber
冷冻室	chilling chamber
箱体	box
电茶壶	electric tea pot
壳体	housing

电磁炉	induction cooker
底座	base
电饭锅,电饭煲	electric cooker
盖板	butt plate
内锅	inside pot
外锅	outside pot
电咖啡机	electric coffee machine
壳体	housing
电烤箱	electric oven
箱体	box
空调	air-conditioner
壁挂式空调	wall-mounted air-conditioner
室内机	internal unit
室外机	outdoor unit
柜式空调	cabinet air-conditioner
室内机	internal unit
室外机	outdoor unit
清洗机	cleaning machine
方箱体	square box
燃气热水器	gas water heater
盖	cover
框,箱	case
燃气灶	gas stove
底座	base

微波炉	microwave oven
门框	door frame
箱体	box
吸油烟机	extractor hood
罩	cover
洗碗机	dishwasher
水桶	bucket
洗衣机	washing machine
滚筒	drum
箱体	box

5.4 五金工具，日用器皿　Hardware Tools, Daily Utensil

餐叉	table fork
餐匙	table spoon
餐刀	table knife
尺	ruler
刀	knife
电工刀	electrician's knife
钢卷尺	steel measuring tape
刮刀	scraper
剪刀	scissor
金属直尺	metal ruler
锯	saw

带锯	band/strap saw
电锯	electric saw
钢锯	steel saw
弓锯	bow saw
鸡尾锯	cocktail saw
锯条	saw blade
手板锯	panel saw
圆锯	rim/ring saw
卷尺	measuring tape
量规	gauge
厚度规	finger gauge
节距规	pitch gauge
内外卡钳	double calliper
盆	basin
碗	bowl
万能角尺	universal angle ruler
油灰刀	putty knife
游标卡尺	vernier caliper
圆规	compasses
直角尺	L-square

06 金属材料　Metal Materials

材料	material
导电材料	conductor material
镀层材料	cladding material
复合材料,合成材料	composite material
工程材料	engineering material
航空材料	aerial material
热轧材料	hot-rolled material
双金属材料	bimetallic material
钢	steel
奥氏体不锈钢	austenitic stainless steel
奥氏体钢	austenitic steel
八角钢	octagon steel
半硬钢	half-hard steel
半镇静钢	semi-killed steel
包层薄钢板	clad sheet steel
包层钢板	clad steel
扁钢,钢条	flat-rolled steel
表面硬化钢	case-hardened steel

不变形钢	non-deforming steel
不锈钢	stainless steel
超高强度钢	super-high-strength steel
沉淀硬化不锈钢	precipitation hardening stainless steel
磁钢	magnetic steel
淬火钢	hardened steel
带钢	band steel
弹簧钢	spring steel
低合金钢	low-alloy steel
低合金高强度钢	low-alloy high-strength steel
低合金工具钢	low-alloy tool steel
低碳钢	low-carbon steel
电机钢	dynamo steel
锭钢	ingot steel
锻钢	forged steel
多元合金钢	complex alloy steel
二元合金钢	binary steel
沸腾钢	boiling steel
复层钢,复合钢	composite steel
高合金钢	high-alloy steel
高级钢	high-grade steel
高级优质钢	extra-fine steel
高强度钢	high-tensile steel

高强度结构钢	high-tensile structural steel
高速钢	quick-cutting/high-speed steel
高速工具钢	high-speed tool steel
高碳钢	high-carbon steel
工具钢	tool steel
工字钢	I-steel
光亮精整钢	bright-finished steel
滚珠轴承钢	ball-bearing steel
过烧钢	burnt steel
合金钢	alloy steel
合金工具钢	alloy tool steel
厚钢板	steel plate
极硬钢	dead-hard steel
结构钢	structural steel
结构合金钢	structural alloy steel
拉光钢	bright-draw steel
冷镦模具钢	cold-heading die steel
冷拉钢	cold-drawn steel
冷轧带钢	cold-strip steel
冷轧钢	cold-rolled steel
六角钢	hexagonal steel
马氏体不锈钢	martensitic stainless steel
耐高温钢	high-temperature steel
耐高温钢板	high-temperature steel sheet

耐磨钢	wear-resisting steel
耐磨合金钢	wear-resisting alloy steel
耐热钢	heat-resisting steel
耐蚀钢	corrosion-resisting steel
耐酸钢	acid-resisting steel
炮钢	gun steel
普通碳素钢	common carbon steel
强烈沸腾钢	wild steel
全淬硬钢	full-hardened steel
热锻模钢	hot die steel
热轧钢	hot-rolled steel
软钢	mild/soft steel
软钢板	mild steel sheet
三元合金钢	ternary alloy steel
深拉钢	deep-drawing steel
深硬化钢	deep-hardening steel
渗碳钢	carburizing/cementation steel
碳素钢	carbon steel
碳素工具钢	carbon tool steel
特高强度钢	extra-high tensile steel
特软钢	extra-soft steel
特殊钢	special steel
特殊合金钢	special alloy steel
条钢	bar steel

铁素体钢	ferrite steel
无磁性钢	non-magnetic steel
矽钢,硅钢	silicon steel
型钢	profile steel
液态加压钢	liquid-compressed steel
易切削钢	free-cutting steel
优质钢	fine steel
镇静钢	killed steel
中碳钢	medium carbon steel
珠光体钢	pearlitic steel
铸钢	cast steel
钢板,板,片	sheet
波纹薄板	corrugaed sheet
薄钢板	light-gauge sheet
薄片	thin sheet
电工钢片	electrical sheet
电机钢片	dynamo steel sheet
镀镍板	nickel plated sheet
镀锌钢皮	galvanized sheet
光亮钢板	bright luster sheet
夹层金属薄板	plywood metal sheet
精轧薄板	planished sheet
蓝光薄钢板	polished blue sheet
冷轧薄钢板	cold-rolled sheet

汽车车体薄板	automobile body sheet
软钢皮	mild steel sheet
酸洗薄钢板	pickled steel sheet
无光薄板	dull-finished sheet
锡片	tin metal sheet
锌片	zinc metal sheet
合金	alloy
α-β 黄铜合金	alpha-beta brass alloy
α-β 钛合金	alpha-beta titanium alloy
α 黄铜合金	alpha brass alloy
α 钛合金	alpha titanium alloy
β 黄铜合金	beta brass alloy
β 钛合金	beta titanium alloy
巴氏合金	babbitt alloy
铂铑合金	platinum-rhodium alloy
超高速切削合金	super-high-speed cutting alloy
磁性合金	magnetic alloy
导电合金	electrical conductivity alloy
低膨胀合金	low-expansion alloy
低熔点合金	low melting alloy
电阻合金	resistance alloy
多相合金	heterogenous alloy
二元合金	binary alloy
非磁性合金	non-magnetic alloy

高导磁率合金	high permeability alloy
高电阻合金	high resistance alloy
高欧姆电阻合金	high ohmic resistance alloy
高熔点合金	high melting point alloy
高塑性合金	high-ductile alloy
高温合金	high temperature alloy
共晶合金	eutectic alloy
共析合金	eutectoid alloy
固溶合金	solid solution alloy
过共析合金	hypereutoctoid alloy
海军黄铜合金	naval brass alloy
合成铸铁合金	synthetic cast-iron alloy
黄铜合金	yellow brass alloy
活性合金	vital alloy
加工硬化合金	work-hardening alloy
精密合金	precision alloy
均质合金	homogeneous alloy
抗磁合金	diamagnetic alloy
可锻铝合金,软铝合金	wrought aluminium alloy
铝合金	aluminum alloy
铝基合金	aluminium base alloy
镁合金	magnesium alloy
耐磨合金	wear-resistant alloy

耐热合金	heat-resisting alloy
耐氧化合金	non-oxidizable alloy
尼莫尼克合金	nimonic alloy
轻合金	light alloy
三元合金	ternary alloy
四元合金	quarternary alloy
特硬工具合金	superhard tool alloy
铁磁性合金	ferromagnetic alloy
铁碳合金	iron-carbon alloy
铜基合金	copper base alloy
透磁合金	permeability alloy
伍德合金	Wood's alloy
易熔合金	fusible alloy
硬合金,硬质合金	hard alloy
硬铝合金	duralumin alloy
有色金属合金	non-ferrous metal alloy
中间合金	master alloy
重合金	heavy alloy
铸造合金	casting alloy
黄铜	brass
海军黄铜	naval brass
金属	metal
半成品金属	semi-finished metal
不锈金属	rustless metal

复合金属	composite metal
贵金属	noble metal
海绵状金属	spongy metal
航空金属	aerial metal
黑色金属	ferrous metal
减磨金属	antifriction metal
耐熔金属	refractory metal
稀土金属	rare-earth metal
稀有金属	rare metal
硬金属	hard metal
有色金属	non-ferrous metal
再用金属	secondary metal
重金属	heavy metal
青铜	bronze
铁	iron
α铁	alpha iron
β铁	Beta iron
δ铁	Delta iron
阿姆科铁,工业纯铁	Armco iron
白口铁	white iron
扁铁	flat iron
扁铁条	band iron
不锈铁	rustless iron
槽钢	channel iron

磁铁	magnet
等边角铁	equal angle iron
锭铁	ingot iron
锻铁	forge iron
方铁	square iron
废铁	scrap/waste iron
高级铸铁	high-grade cast iron
高强度铸铁	high-strength cast iron
工字铁	double iron
合金铁	alloyed iron
合金铸铁	alloy cast iron
黑钢皮	black sheet iron
灰口铁	gray iron
角铁	corner iron
可锻铸铁	malleable cast iron
六角铁	hexagon iron
普通铸铁	common iron
生铁	pig iron
熟铁	wrought iron
铁板	iron plate
铁皮	iron sheet
铁条	bar iron
型铁	fashioned/figured/profiled/sectional/shape iron

延性铁	ductile iron
优质铸铁	high-quality cast iron
圆铁	rod/round iron
轧制铁皮	rolled sheet iron
铸铁	cast iron

07 钢锭冶炼、结构及缺陷　　Ingot Smelting, Structure and Defects

白点,鳞片	fish scale/flake
白点,缩孔	fish eye
斑点,斑痕	spot
斑点	blemish
鲍曼硫印	Baumann print
层,薄层,薄膜	layer
粗大晶粒层	coarse grained layer
细晶粒层	fine grained layer
沉积,沉积物	deposit
除气,脱气	degassing
锭,钢锭,铸锭	ingot
八边形钢锭,八角形钢锭	octagonal ingot
扁钢锭,扁锭	flat-shaped ingot
薄皮钢锭	thin-skinned ingot
锭模	ingot mould
锭身	ingot body

短粗钢锭	short-gross ingot
短钢锭	squat ingot
锻造钢锭	forging ingot
多边形钢锭,多角形钢锭	polygonal ingot
多锥度钢锭	multicone ingot
沸腾钢锭	rimming ingot
钢锭尺寸	ingot size
钢锭底部	ingot bottom
钢锭规格	ingot standard
钢锭缺陷	ingot defect
钢锭外形	ingot configuration
钢锭形式	ingot form
钢锭形状	ingot shape
钢锭种类	ingot kind
钢锭重量	ingot weight
钢锭锥度	ingot cone
过烧钢锭	burned ingot
空心钢锭	hollow ingot
普通钢锭	common ingot
无缩孔钢锭	balanced ingot
细长钢锭	thin-long ingot
有凹口的钢锭	notched ingot
有气孔钢锭	blown ingot

圆锭	round ingot
断层	fault
闭断层	closed fault
叠断层	multiple fault
断隔断层	gaping fault
分支断层	distributive fault
共轭断层	conjugated fault
横断层	cross fault
挤压断层	compression fault
交错断层	intersecting fault
阶状断层	echelon fault
裂缝断层	fissure fault
倾斜断层	dip fault
倾斜断层	inclined fault
倾斜滑断层	dip-slip fault
曲断层	cured fault
斜断层	diagonal fault
圆柱状断层	cylindrical fault
褶皱断层	folded fault
重力断层	gravity fault
纵断层	longitudinal fault
多孔性,疏松	porosity
飞溅	splash
夹层	lamination

夹渣,夹杂物	dirt
溅末	spatter
浇补	casting-on
浇口	sprue
浇注	pouring
补浇注	back pouring
上注法	top pouring
下注法,底注法	bottom pouring
结疤	scab
晶体,晶粒	crystal
粗大等轴晶体	coarse grained equiaxia crystal
等侧轴晶体	isodiametric crystal
二轴晶体,双轴晶体	biaxial crystal
各向异性晶体	anisotropic crystal
立方晶体	cubic crystal
树枝晶粒	dentritic crystal
树状晶体	arborescent crystal
针状晶体	acicular crystal
中心对称晶体	centrosymmectrical crystal
柱状晶体	columnar crystal
孔隙,气孔	pore
冷焊	cold welding
裂缝,裂纹,裂痕	crack
表面裂缝	face crack

初裂	incipient crack
淬火裂纹	hardening/quenching crack
锻缝	forge crack
发裂,发纹,微缝	hair crack
放射裂纹	radiating crack
横向不规则裂纹	pull crack
滑裂,层裂	slip crack
加热裂缝	fire crack
剪切裂纹	shearing crack
毛细裂缝	capillary crack
内裂	internal crack
热处理裂纹	heat treatment crack
热裂	hot crack
收缩裂纹	check/contraction/shrinkage crack
水淬裂纹	water crack
轧制裂纹	rolling crack
皱裂	crow-foot crack
冒口,切头	head metal
冒口,切头,废料	lost head
冒口	riser
偏析	segregation
表面偏析	surface segregation
反偏析	inverse segregation
负偏析	negative segregation

宏观偏析	macro segregation
晶间偏析	intercrystalline segregation
局部偏析	spot segregation
磷偏析	phosphorus segregation
区域偏析	zone segregation
缩孔偏析	pipe segregation
网状偏析	cellular segregation
微观偏析	micro segregation
正偏析	positive segregation
气孔	air hole
切头（冒口部分）	crop end
氢脆	hydrogen brittleness
区，区域	zone
粗晶粒区	coarse grained zone
偏析区	segregation zone
倾斜树枝晶区	inclined dentritic zone
柱状晶区	columnar grained zone
缺陷	defect
浇注缺陷	pouring defect
晶格缺陷	lattice defect
晶体缺陷	crystal defect
质量缺陷	mass defect
熔炼	melting
磁性电炉熔炼	magnetic electric melting

电弧熔炼	arc melting
坩埚钢熔炼	crucible steel melting
还原熔炼	reduction melting
炼钢	steel melting
区域熔炼	zone melting
氧化熔炼	oxidizing melting
真空电弧熔炼	vacuum arc melting
真空熔炼	vacuum melting
砂眼	sand hole
收缩,缩孔	shrinkage
残余收缩	residual shrinkage
内部收缩	interior shrinkage
凝固收缩	setting shrinkage
凝缩	solidification shrinkage
体积收缩	volume shrinkage
温差收缩	thermal shrinkage
收缩	contraction
侧收缩	lateral contraction
端收缩	end contraction
金属液态收缩	metal liquid contraction
局部收缩	local contraction
重力收缩	gravitational contraction
撕裂擦伤	tearing
缩孔	shrinkage cavity

碳,碳精棒	carbon
金属芯碳极	metal-cored carbon
碳精电极,碳极	baked carbon
铁渣,浮渣	dross
精炼炉渣	refinery dross
硫化矿渣	pyrite dross
阳化物浮渣	oxide dross
脱壳	decortication
未铸满,缺肉	misrun
杂质,夹杂物	inclusion
非金属夹杂物	non-metallic inclusion
夹渣	slag inclusion
内在夹杂物	internal inclusion
外来夹杂物	foreign inclusion
氧化夹杂物	dross inclusion
杂质	impurity
机械杂质	mechanical impurity
金属杂质	metal impurity
折叠	lap
褶皱,折叠	fold
表层折叠	mantle fold
断层折叠	fault fold
阶状折叠	step fold
流线折叠	flow fold

弯曲折叠	flexural fold
重褶折叠	refolded fold
铸造,铸件	casting
薄壁铸件	thin-section casting
薄壳铸造	shell mold casting
厚壁铸件	thick-walled casting
精密铸件	precision casting
精确离心铸造	true centrifugal casting
块铸	block casting
蜡模铸造,熔模铸造	investment casting
冷硬铸法	case hardened/chilled casting
离心铸法	centrifugal/rotary casting
连续浇铸	continuous casting
气孔铸件	blown casting
熔模精密铸造	precision investment casting
砂箱铸造	flask casting
上铸法	top casting
双金属铸件	composite casting
下铸法,底铸法	bottom/uphill casting
压力浇铸	pressure casting
压铸	compression casting
硬模铸造	die casting
有色金属铸件	non-ferrous casting
真空浇铸	vacuum casting

真空压力铸造	vacuum die casting
整体铸法	integral casting
直接铸造(法)	direct casting
中心浇铸(法)	centre casting
重力铸造	gravity die casting
锥度	taper
倒锥度	back taper

08 下料及检验 Cropping and Inspection

棒料剪断机	bar cropping machine
刀片	blade
固定刀片	stationary blade
活动刀片	moving blade
下刀片	bottom blade
刀刃	knife edge
机械剪切机	mechanical shear machine
检验,检查	inspection
X射线检查	X-ray inspection
不定期检查,临时检查	casual inspection
采样检查	sampling inspection
常规检查	routine inspection
超声波检查	ultrasonic inspection
抽查	random inspection
抽样检查	curtailed inspection
磁粉检查	magnaflux/magnetic powder inspection

磁性检查	magnetic inspection
磁性荧光法检查	fluorescent magnetic particle inspection
电子检查	electronic inspection
定期检查	periodic/regular inspection
光电检查	photoelectric inspection
光谱检查	spectral inspection
红外线检查	infrared inspection
火花检查	spark inspection
年检	annual inspection
日检查	daily inspection
外部检查,肉眼检查,直观检查	visual inspection
外观检查	outer inspection
完工检查	finish-turn inspection
无损探伤,非破坏性检查	non-destructive inspection
质量检查	quality inspection
剪板机	shearing machine
剪切,截短	cropping
半热剪切	half-hot cropping
高速剪切	high speed cropping
回转剪切	rotary cropping
精密剪切	precision cropping

径向剪切	radial cropping
冷剪切	cold cropping
倾斜剪切	tilted cropping
热剪切	hot cropping
双面剪切	double cropping
轴向加压剪切	cropping with axial compression

截断　　　　　　　　　　cut-out
锯,锯切　　　　　　　　saw

带锯	endless saw
电机械锯切	electro-mechanical saw
电锯	electric motor saw
弓形锯	bow saw
金钢砂锯	carborundum saw
摩擦锯	friction saw
片锯	blade saw
台锯	bench saw
圆锯	circular/disk saw

评定,测定,鉴别　　　　evaluation
切开　　　　　　　　　　cut-off

　砂轮切割　　　　　　　abrasive cut-off

试验,检验,检查　　　　test

埃里克森试验	Erichsen test
百分比试验	percent test
摆冲试验	chary test

摆冲试验	pendulum test
保证试验,认可试验	warranty test
比较试验	comparison test
不损坏试验	non-destructive test
部分试验	detail test
部分系统试验	partial system test
材料成分化验试验	material composition test
产品质量检查	production quality test
超声波试验	ultrasonic test
超速试验	overspeed test
冲击韧性试验	toughness test
冲击试验,动力试验	impact/shock test
冲击硬度试验	impact hardness test
冲弯试验	impact bend test
冲压试验	punching test
抽样试验	pick test
锤击试验,可锻性试验	forge test
单件试验	single test
弹性试验	elasticity test
点滴定性试验	spot quality test
点滴试验	spot test
电火花试验	spark test
顶断试验	bulging test

定量试验	quantitative test
定性试验	qualitative test
动力试验	dynamic/power test
断裂试验	break test
额定负载试验	running test
反跳硬度试验	rebound hardness test
腐蚀试验	corrosion test
高温冲击试验	high temperature test
工厂试验	shop test
过载试验,超负荷试验	overload test
合格试验	qualification test
划痕试验	scratch test
回弹试验,回跳试验	rebound test
回转试验	circle test
火花鉴别法	pellet test
机械试验	mechanical test
加热试验	heating test
加载试验,负载试验	load test
监查试验	monitoring test
剪刀试验	shear test
金相试验	metallographic test
静态试验	static test
抗拉试验	tensile test

抗压试验，压缩试验	compression test
空载试验，无载试验	no-load test
空转试验	motoring test
快速试验，短时试验	short time test
拉断试验	break-down test
拉伸试验	elongation test
冷弯试验	cold beading test
例行实验室试验	routine laboratory test
洛氏硬度试验	Rockwell hardness test
落锤试验，冲击试验	drop test
脉冲试验	impulse test
满载试验，重载试验	heavy test
模拟试验	simulation test
模型试验	model test
耐冲试验	impact endurance test
挠性试验	flexing test
扭转试验	torsion/twisting test
疲劳试验	fatigue test
破坏试验	destruction test
破损试验	crash test
强度试验	strength test
切口棒试验	notch bar test
热弯曲试验	hot-bending test
深拉试验	cup-drawing test

生产定型试验	production-type test
生产试验，产品检验	production test
试件试验	specimen test
寿命试验	life test
水力试验	hydraulic test
斯维夫脱拉深试验	Swift cupping test
探针试验	probe test
弯矩试验	moment test
弯曲试验	bend test
稳定性试验	stability test
涡电流试验	eddy current test
显微镜试验	micrographic test
行车试验	road test
压杯试验，拉伸试验	cup(ping) test
压力试验，加压试验	pressure test
压碎试验	crushing test
液压试验，水试验	water test
伊佐德冲击试验	Izod impact test
应力试验	stress test
硬度试验	hardness test
振动试验	vibration test
正式试验	official test
直观试验	macroscopic test
重复加载试验	repeated-load test

周期试验	periodic test
下料，切割，剪切	shear
侧剪	lateral shear
垂直剪	vertical shear
纯剪	pure shear
单剪	simple shear
电火花切割	electro spark shear
端面剪切	end shear
横剪	transverse shear
两面剪	two-side shear
摩擦锯切割	shear of friction saw
气割	gas shear
砂轮切割	shear of abrasive wheel
双面剪切	double shear
相对剪切	relative shear
纵剪	longitudinal shear
液压剪切机	hydraulic shear machine

09 金属加热及加热炉　Metal Heating and Furnaces

9.1 加热　Heating

保温	soaking
保温时间	hold time
锻造温度范围	forging temperature interval/zone
过热	overheating
过烧	burning
火次	heating number
火色	temperature colour
加热	heating
电加热	electric heating
电阻加热	ohmic heating
电阻加热	resistance heating
分段加热	step heating
辐射加热	radiation heating
感应加热	induction heating
火焰加热	flame heating
加热规范	heating specification

加热规则	heating rule
加热曲线	heating curve
加热时间	heating time
加热速度	heating rate/velocity
加热温度	heating temperature
加热制度	heating system
接触加热	contact heating
局部加热	local heating
快速加热	rapid heating
少无氧化加热	scale-less heating
无氧化加热	non-oxidation heating
盐浴炉加热	salt-bath heating
炉温	furnace temperature
烧损,火耗	heating scale loss
脱碳	decarburation
表面脱碳	skin/surface decarburation
温度头	temperature head
氧化	oxidation
不完全氧化	incomplete oxidation
缓慢氧化	slow oxidation
剧烈氧化	vigorous oxidation
轻度氧化	mild oxidation
氧化铁皮	scale/oxide skin
氧化物	oxide

氧化层	oxide layer
氧化膜	oxide film
氧化皮	oxide skin
装炉温度	charging temperature

9.2 加热炉　　　Furnace

标准燃料	standard fuel
测温笔	tempilstick
二次空气	secondary air
高温仪	pyroscope
鼓风机	blower
光学高温计	optical pyrometer
烘炉	drying furnace
换热器,再生器	regenerator
逆流式热交换器	contraflow regenerator
旋转式回热器	rotary regenerator
加热炉	furnace
敞焰少无氧化加热炉	direct-fired scale-less furnace
地炉	pit furnace
电炉	electric furnace
电阻炉	electric furnace of resistance type
高频感应电炉	high-frequency induction furnace
贯通式炉	passing-type furnace

火焰炉,反射炉	flame furnace
精炼炉	improving furnace
开隙式炉	slot-type furnace
连续炉	continuous furnace
炼钢炉	steel-melting furnace
煤气炉	gas furnace
烧结炉	sintering furnace
室式炉	batch-type furnace
手锻炉	smith forge furnace
双膛炉	double furnace
台车式炉	car-bottom hearth furnace
推杆式炉(半连续炉)	pusher furnace
无氧化加热炉	less oxidation heating furnace
矽碳棒电阻炉	silicon carbide rod furnace
箱式炉	box furnace
盐浴炉	salt bath furnace
真空电弧炉	vacuum arc furnace
真空精炼炉	vacuum fining furnace
振底炉	shock bottom furnace
转壁炉	rotary wall furnace
转底炉	rotary hearth furnace
焦炭	coke
空气消耗系数	coefficient of air consumption

理论燃烧温度	theoretical combustion temperature
炉算强度	intensity of grate
炉底强度	intensity of hearth
炉底热强度	heat intensity of hearth
炉温系数	temperature coefficient
炉子热效率	calorific effect of furnace
炉子生产率	productivity of furnace
煤	coal
层状煤	slabby coal
肥煤	rich coal
炼焦煤	coking coal
劣质煤	fault coal
煤粉	powder coal
碎煤	rice coal
无烟煤	blind coal
无烟煤	glance coal
烟煤	bituminous coal
原煤	raw/rough coal
木炭,炭	char
耐火材料	refractory
隔热耐火材料	insulating refractory
碱性耐火材料	basic refractory
酸性耐火材料	acid refractory
中性耐火材料	neutral refractory

黏粒,黏土	clay
铝质黏土	bauxite clay
耐火黏土	fire/refractory clay
膨润土	bentonitic clay
轻质黏粒	hydrogen clay
酸性黏土	acid clay
细耐火泥	finely-ground fire clay
喷嘴	nozzle
多孔喷嘴	multi-jet nozzle
可调喷嘴	variable nozzle
平口喷嘴	smooth nozzle
收敛形喷嘴	constrictor nozzle
油喷嘴	oil nozzle
针形喷嘴	needle nozzle
锥形喷嘴	conical jet nozzle
气,气体,煤气	gas
半煤气	semi-coal gas
半水煤气	semi-water gas
城市煤气	city gas
惰性气体	inert gas
二氧化碳	carbon dioxide
发生炉煤气	producer gas
非理想气体	imperfect gas
废气	burnt/end/off/waste gas

富氧煤气	oxygen-enriched coal gas
高炉煤气	blast furnace gas
可燃气体	combustible gas
冷煤气	cold coal gas
理想气体	ideal gas
炉气	burner gas
煤气	coal gas
燃烧气体	inflamed gas
热煤气	hot coal gas
石油气	oil gas
天然气	rock gas
易燃气体	inflammable gas
区,区域,范围	zone
饱和区域	zone of saturation
保护区	zone of protection
火焰区	zone of flame
加热区	zone of heating
均热段	soaking zone
燃烧段	burning zone
热影响区	heat-affected zone
全辐射高温计	all-radiation pyrometer
燃料,燃油	fuel
参考燃料,标准燃料	reference fuel
超级燃料	superior fuel

低比重燃料	low-gravity fuel
低挥发性燃料	low-volatility fuel
废燃料	spent fuel
粉状燃料	pulverized fuel
高能燃料	high-energy fuel
混合燃料	blended fuel
块状燃料	clumped fuel
流体燃料	fluid fuel
普通燃料	conventional fuel
轻燃料	light fuel
无烟燃料	smokeless fuel
液化气体燃料	liquid gas fuel
原始燃料	parent fuel
再生燃料	recycled fuel
燃料单耗(煤铁比)	specific consumption of fuel (ratio between coal and iron)
燃料发热值	calorific value of fuel
燃料利用系数	coefficient of fuel utilization
热电高温计	thermoelectric pyrometer
热电偶	thermocouple
双金属热电偶	bimetallic thermocouple
热平衡	heat balance
烧嘴,燃烧器	burner
低压鼓风烧嘴	low-pressure blast burner

低压烧嘴	low-pressure burner
高速烧嘴	high velocity burner
高压烧嘴	high-pressure burner
喷射式烧嘴	injection burner
平焰烧嘴	flat-flame burner
转杯烧嘴	rotary cup burner
自身预热烧嘴	self-recuperative burner
石棉	asbestos
石棉粉	flaked/pulverized asbestos
石棉片	sheet asbestos
示温涂料	thermocolor
室,燃烧室	chamber
反应室	reaction chamber
分燃室	separate combustion chamber
还原室	reduction chamber
开式室	open chamber
喷射室	nozzle chamber
预热室	preheating chamber
真空室	vacuum chamber
温度计	thermometer
烟,冒烟	smoke
烟囱	smoke-tube
烟道	chimney
一次空气	primary air

油,石油	oil
柴油	diesel oil
锭子油	bobbin oil
机油	engine oil
煤油	kerosene oil
轻燃料油	light fuel oil
原油	crude oil
重柴油	heavy diesel oil
重油	heavy oil
预热	preheat
预热器	preheater
砖	brick
薄边砖	feather-edge brick
超级耐火砖	super-duty brick
多孔砖	cork brick
高级耐火砖	high refractory brick
拱顶砖	crown brick
弧形砖	circle brick
火泥砖	chamber brick
角砌砖	edge brick
绝热砖	cellinsulate brick
绝缘砖	insulating brick
空心砖	cavity brick
炉底砖	bottom brick

耐火砖	refractory brick
耐酸砖	acid-proof brick
内八角砖	internal octagon brick
黏土砖	clay brick
曲面砖	curved brick
梯形砖	trapezoidal brick
外八角砖	external octagon brick
习用砖	conventional brick
粘合砖	cementitious brick
装料方式	charging method

10　锻压设备及装置　Forging-Stamping Equipment and Device

10.1　锻造设备及零件　Forging Equipment and Parts

泵	pump
齿轮泵	gear pump
活塞泵	plunger pump
离心泵	centrifugal pump
螺旋泵	helical pump
液力泵	hydraulic pump
油泵	oil pump
扁心齿轮传动	eccentric gear drive
超负荷保险装置	overload protector
齿轮	gear
充液行程	charge stroke
锤,汽锤	hammer
板簧锤	laminated spring hammer
带式落锤	strap hammer
单臂锤	overhanging hammer

单动汽锤	single acting steam hammer
单柱锤	single frame/C-frame hammer
单柱汽锤	overhanging-type steam hammer
单柱式锤	single flame hammer
单作用锤	drop action hammer
弹簧锤	spring hammer
低速内燃锤	low-speed petroforge hammer
电锤	electric hammer
动力（落）锤	power (drop) hammer
锻锤	blacksmith's/forging hammer
锻工平锤	blacksmith's flat hammer
对击锤	counterblow hammer
杠杆锤	helve/lever hammer
高速锤	high energy rate forging hammer
高速锤	high velocity hammer
高速液压锤	high-speed hydraulic hammer
拱式自由锻锤	double flame forging hammer
机械锤	machine hammer
夹板锤	jump/board drop hammer
开坯锻锤	cogging hammer
空气锤	air/pneumatic hammer
空气落锤	air drop hammer
链条落锤	chain drop hammer
龙门汽锤	arch type steam hammer

铆钉锤	cup-shaped hammer
模锻锤	die forging/swage hammer
摩擦锤	friction hammer
摩擦辊轮落锤	friction roll drop hammer
皮带锤	belt drop hammer
桥式锤	bridge type hammer
手锤	hand hammer
数控全液压模锻锤	die forging hammer of full hydraulic with numerical control
双缸汽锤	compound hammer
双柱式锤	double flame hammer
双作用锤	double action hammer
凸轮锤	cam hammer
校正锤	straightening hammer
压气锤	compressed hammer
液压锤	hydraulic hammer
液压螺旋锤	hydraulic screw hammer
圆盘锤	disk hammer
圆盘摩擦落锤	friction-board hammer
砧座锤	anvil hammer
蒸汽锤	steam hammer
蒸汽-空气模锻锤	steam-air die forging hammer
蒸汽-空气自由锻锤	steam-air forging hammer

重力落锤,夹板锤	gravity drop hammer
自由锻锤	smith/forging hammer
锤头	hammer ram
寸动	adjusting microinching
打击	blow
打击能量	blow energy
打击速度	blow speed
打击效率	blow efficiency
单次行程	single stroke
导板,导轨	guide
对角斜置导轨	diagonal guide
底座,支座	base
地基	foundation
电磁离合器	electromagnetic clutch
垫木	sole timber
垫圈	wasjer
吊钳	hanging tongs
顶出力	ejecting force
阀	valve
安全阀	safety valve
给水阀	feed water valve
控制阀	control valve
四通阀	four-way valve
溢水阀	flood valve

止回阀,单向阀	back valve
飞轮	fly wheel
封闭高度调节量	shut height adjustment
封闭高度调节装置	adjusting device for shut height
刚性离合器	rigid clutch
工作台	table/bed
工作台尺寸	bed dimension/table size
工作行程	working stroke
工作行程速度	speed of working stroke
公称压力	capacity
公称压力行程	nominal working stroke
喉深	throat depth
滑块	slide/ram
回程	return stroke
回程速度	speed of return stroke
活动工作台行程	travel of moving bolster
机器	machine
摆动辊压机	swing rolling machine
剥皮机	peeling machine
冲孔机	piercing machine
冲切机	dinking machine
穿孔机	punching/perforating machine
等离子切割机	plasma cutting machine
电镦机	electric upsetting machine

镦锻机	upsetting machine
钢板矫平机	plate-levelling machine
钢板精整机	plate-straightening machine
辊锻机	roll forming machine
辊式矫直机	roller straightening machine
横轧机	transverse rolling machine
回转剪切机	revolving shearing machine
火焰切割机	flame cut-off machine
激光切割机	laser cutting machine
挤压机	extruding machine
剪板机	plate cutting machine
精密锻造机，精锻机	precision forging machine
径向锻轴机	radial forging machine
径向精密锻造机	radial precision forging machine
卷板机	plate rolling machine
扩管机	pipe expanding machine
拉拔机	drawing machine
拉形机	stretching machine
立式卷板机	vertical plate rolling machine
螺旋横轧机	screw transverse rolling machine
模锻机	swaging machine
喷丸机	shot blast chamber machine
平锻机，卧式锻机	horizontal forging machine
气动冲床	pneumatic punching machine

切边机	trimming machine
切断机	cutting machine
切管机	pipe cutting machine
三辊卷板机	three-roller plate rolling machine
砂轮切割机	abrasive cut-off machine
上料机	appending machine
水切割机	water machine
四辊卷板机	four-roller plate rolling machine
送料机	feeder machine
弯板机	plate bending machine
弯边机,卷边机	beading machine
弯管机	pipe bending machine
弯曲机	bending machine
卧式弯板机	holizontal plate-bending machine
楔横轧机	wedge rolling machine
旋压机	spinning machine
旋转模锻机	rotary swaging machine
压纹机	embossing machine
液压冲床	hydraulic punching machine
液压弯板机	hydraulic plate-bending machine
折边机	closing/cramp folding/crimping machine
机身	frame
夹持力矩	clamping torque

结构	structure
焊接结构	welding structure
夹层结构	sandwich structure
整体结构	integral structure
铸造结构	cast structure
空行程	idle stroke
快速行程	accelerated stroke
立柱	upright
立柱净空距	clear distance between uprights
连续行程	continuous operation
梁	beam
合成梁	compound beam
横梁	cross beam
桥式梁	bridge beam
上横梁	top cross beam
下横梁	bottom cross beam
悬臂梁	cantilever beam
圆梁	circular beam
柱形梁	column beam
组合梁	combination beam
螺栓	bolt
地脚螺栓	long bolt
调整螺栓	adjustable bolt
防松螺栓	check bolt

固定螺栓	fixing bolt
夹紧螺栓	clamp bolt
紧固螺栓	trip bolt
拉紧螺栓	maneton/pinch bolt
连接螺栓	attachment bolt
支撑螺栓	carrying bolt
止动螺栓	catch bolt
落下部分	falling part
每分钟打击次数	blows per minute
每分钟精整快锻次数	numbers of finishing strokes
模具高度调节量	die set height adjustment
皮带	belt
V-形带	V-type belt
帆布带	canvas belt
交叉带	cross belt
链带	chain belt
牛皮带	leather belt
三角皮带	triangle belt
平衡装置	balancing device
气垫	cushion/air cushion
汽缸	steam cylinder
汽套	steam jacket
曲柄连杆机构	crank link mechanism
曲柄肘杆机构	toggle mechanism

上死点	upper dead point
上行程	upstroke
双边传动	double sided gear drive
双速离合器	two speed clutch
外滑块停歇时间	outer slide dwell time
温度监控	temperature monitor
下死点	lower dead point
下行程	down-stroke
楔,楔块	wedge
液压气垫	hydropneumatic cushion
圆盘摩擦离合器	disk friction clutch
轧钢机	mill
穿孔机	piercing mill
粗轧机	big mill
钢坯轧机	billet mill
精轧机	planishing mill
开坯机,初轧机	blooming mill
热轧机	hot-rolling mill
三辊式开坯机	three-high blooming mill
斜轧穿孔机	cone piercing mill
砧座	anvil bed
制动力矩	brake torque
制动器	brake
重量	weight

净重	net weight
空重，皮重	bar/tare weight
落下部分重量	droping weight
实重，真重	true weight
原重	original weight
总重	total weight
主柱塞	main ram
柱塞	plunger piston
转数表	tachometer
最大打料行程	maximum knock-out stroke
最大封闭高度	maximum shut height
最大偏心距	maximum eccentricity
最大行程	maximum stroke
最大装模高度	maximum die set height

10.2 冲压设备及零件　Stamping Equipment and Parts

操纵台	benchboard
传感器	sensor
调节器	regulator
分配器	distributor
功率计	dynamometer
滚剪机	plate rotary shear
缓冲器	buffer

换向器	deffector
记录器	recorder
剪板机	plate shear
开卷校平剪切机	uncoil-flattening shears
离合器	clutch
两面切边机	two side shear
漏斗,料箱	hopper
平衡器	balancer
示波器,示波仪	oscilloscope
台	desk
测试台	test desk
观察台	observation desk
中心控制台	central control desk
温度监测器	temperature monitor
选择器	chooser
压力表	manometer
压力机,冲床	press
CNC压力机	computer numerical control press
闭式压力机	straight side press
层压压力机	laminating press
缠绕式液压机	wire-wound hydraulic press
超高压液压机	superhigh pressure hydraulic press
成形压力机	forming press
冲孔压力机	punching/piercing press

冲孔压力机	puncturing press
打包压力机	packaging press
单动压力机	single action press
电磁螺旋压力机	electroc-magnet screw press
锻压机	forging press
锻造水压机	forging pump press
镦粗机	heading press
多冲模冲床,连续冲床	multiple-die press
多缸式液压机	multi-cylinder hydraulic press
多工位冷镦压力机	cold header press with multi-stations
多工位自动压力机	multistation transfer press
多向模锻压机	multi-cored forging press
多柱式液压机	multi-column hydraulic press
高速压力机	high speed press
机械压力机	mechanical press
脚踏板压力机	foot press
脚踏压力机	kick press
精冲压力机	precision blanking press
精压压力机	coining press
精整压力机	sizing press
开式压力机	C-frame press
可斜式压力机	inclinable press

快速锻造水压机	high speed forging hydraulic press
扩孔机	staving press
冷镦机	cold header press
冷挤压力机	cold extrusion press
立柱式压力机	column press
螺旋压力机	fly/screw press
模锻机	stamping press
模锻压力机	drop press
摩擦螺旋压力机	friction screw press
摩擦压力机	friction press
偏心压力机	cam press
偏心压力机	cam/eccentric press
汽力锻压机	steam power forging press
切边压力机	trimming press
切断压力机	cutting-out press
曲轴压力机	crank press
曲轴压力机	crank press
热模锻压力机	hot die press
热模锻压力机	hot drop press
深拉压力机	cupping press
深拉压力机	drawing press
手压机	hand-power press
双动压力机	double action press

双曲柄压力机	double-crank press
双柱压力机	double-arm press
水压机,液压机	hydraulic press
伺服压力机	servo press
台式压力机	bench press
卧式挤压机	extrusion press of horizontal type
卧式金属挤压机	horizontal mechanical extrusion press
下冲式压力机	down stroke hydraulic press
下滑块式拉深压力机	bottom slide drawing press
橡皮模压力机	rubber die press
压平机,矫正机,压直机	gap press
压平机	gagging press
压弯机	bending press
压印机	embossing press
液压螺旋压力机	hydraulic screw press
硬币压力机	mining press
预成型压力机	prefilling press
折边压力机	flanging press
蒸汽液压机	steam hydraulic press
转塔式冲床	turret punch press
自动冲槽机	automatic piercing-slot press
自动压力机	automatic press

油压剪床	oil pressure shears
振动机	nibbling shear
指示器	indicator
制动器	brake

11 锻造　Forging

11.1　自由锻　　Open Die Forging/Free Forging

RR 锻造	RR forging
TR 锻造	TR forging
α+β 锻造	α+β forging
β 锻造	β forging
凹挡	recess
拔长	drawing/stretching
拔长比	ratio of draw
拔长效率	drawing efficiency
长径比	length-diameter ratio
冲孔,冲头	punch
单面冲孔	dressed one side punch
空心冲头	hollow punch
扩孔冲头	expansive punch
实心冲头	solid punch
双面冲孔	dressed two side punch
冲孔	punching/piercing
冲头扩孔	expanding with a punch

锤头	peen
横向锤头	cross peen
圆形锤头	ball peen
粗锻，初锻	rough forge
错开	stagger
错移	offset
倒棱	chamfer
锻件公称尺寸	nominal dimension of forging
锻接，锻焊	forging welding
锻造比	forging ratio
锻造公差	forging tolerance
镦粗	compression/upsetting
镦粗比	ratio of upset
剁刀	chopper
法兰	flange
反复拔长	repeatedly drawing
钢坯	billet
薄板坯	slab billet
方钢坯	square billet
钢板坯	sheet billet
空心钢坯	hollow billet
圆形钢坯	round billet
轧制钢坯	rolling billet
高径比	ratio of height to diameter

滚圆	rolling
黑皮	black body
夹钳	tongs
冲孔钳	punching tongs
带箍的锻工钳	sliding tongs
锻工钳	forge tongs
方口钳	square mouth tongs
钢锭开坯钳	stripping tongs
夹锭钳	ingot tongs
开口钳	open mouth tongs
空心钳	hollow bit tongs
平口钳	flat tongs
起重钳	hoisting tongs
全自动夹钳	full automatic tongs
十字空心钳	double hollow bit tongs
双口锻工钳	double pick-up tongs
弯咀钳	angle jaw tongs
万能钳	universal tongs
圆边口钳	round side mouth tongs
圆口钳	round mouth tongs
錾钳	chisel tongs
夹子	clip
精锻	finish forge
局部镦粗	local upsetting

卡套	clipped socket
开坯	breakdown
挤压开坯	extrusion breakdown
开坯模	breakdown die
扩孔	expanding
漏孔	punching on a supporting
螺旋拔长	screw drawing
马杠扩孔	saddle forging
马架	saddle
毛坯,坯料	blank
板坯	plate blank
半成品	half blank
扁坯	flat blank
齿轮毛坯	gear blank
锭坯	ingot blank
锻造毛坯	forging blank
铸造毛坯	casting blank
铆镦	mushroom upsetting
扭曲,翘曲	warping
扭转	twisting/torsion
平砧间镦粗	compression between flat platens
撬棒	prize
切槽刀	wedged knife
切割	cutting/incision

送进量	feed
台阶	step
铁砧	anvil
凸肩	hub
弯曲	bending
校正	correction
楔块扩孔	expanding with wedge blocks
芯轴,芯棒	mandrel
芯轴拔长	drawing out with the core bar
压扁	flatten
压痕	indentation
压肩	necking
压钳口	pressure clamp
压钳口	tongs hold
压缩量	compression amount
压下量	reduction amount
绝对压下量	absolute reduction amount
相对压下量	relative reduction amount
延伸系数	coefficient of stretching
液压胀形扩孔	hydraulic expanding
余块	excess metal
余面	lap
砧块	block
V形砧,缺口砧	V-shaped anvil block

凹形砧块	concave anvil block
宽砧	wide anvil block
平砧	flat anvil block
凸形砧块	convex block
型砧	swage block
窄砧	narrow anvil block
中心压实法	central compaction method

11.2 模锻　　　　Die Forging

冲连皮	punching the wad
顶镦	heading/upsetting
工序,工步	process
拔长工步	drawing process
辊挤工步	roller process
辊压工步	edging/edge rolling process
卡细工步	reduction process
卡压工步	fullering process
弯曲工步	bending process
压扁工步	squash process
制坯工序	preforming process
精压	coining
平面精压	flat coining
体积精压	bulk coining
模锻	die forging

闭式模锻	closed/non-flash die forging
多向模锻	multi-ram die forging
分段模锻	sectional die forging
开式模锻	open die forging
小飞边模锻	die forging with small flash
切边	trimming
校正	sizing
冷校正	cold sizing
热校正	hot sizing
预锻	blocking/preforging
终锻	finish forging

12　冲压　Stamping

成形	forming
爆炸成形	explosive forming
超声波成形	supersonic wave forming
超塑性成形	super-plasticity forming
电磁成形	electro-magnetic forming
电液成形	electro-hydraulic forming
滚压成形	roll forming
激光冲击成形	laser lash forming
静液压成形	static hydraulic forming
卷拉成形	stretch-wrap forming
拉张-拉深成形	stretch draw forming
气压成形	pneumatic forming
软模成形	flexible die forming
橡皮成形	rubber pad forming
旋压成形	spin forming
液压-橡皮囊成形	rubber-diaphragm forming
液压成形	hydraulic forming
真空成形	vacuum forming

冲孔	punching/piercing
翻边	flanging
翻孔	hole flanging/plunging
废料	scrap
工艺废料	technological scrap
结构废料	structural scrap
分离	cut
辊弯	roll bending
矫直	straightening
卷圆	curling
扩口	flaring/expanding
拉深	drawing
拉弯	stretch bending
拉形	stretch
落料	blanking
扭曲	twisting
剖切	parting
起伏	embossing
切断	cut-off
切口	lancing
缩口	necking
弯曲	bending
校平	flattening
旋压	spinning

变薄旋压,强力旋压	power spinning
锥形变薄旋压	cone power spinning
筒形变薄旋压	tube power spinning
剥皮旋压	spinning with chip forming
错距旋压	stepped spinning
反旋压	backward/indirect spinning
滚珠旋压	ball spinning
加热旋压	hot spinning
精整旋压	planishing/smoothing spinning
扩径旋压	expanding/bulging spinning
内旋压	internal spinning
普通旋压	conventional spinning
缩径旋压	necking/spindown spinning
张力旋压	spinning with tension
正旋压	forward/direct spinning
压印	coining
胀形	bulging
折边	folding
整形	sizing

13 锻压工艺 Forging-Stamping Technology

13.1 模锻工艺　　Die Forging Technology

边,缘	rim
仓部,料仓	bin
槽,沟	slot
冲击数	number of strokes
单击镦锻	single-blow heading
断面减缩率	area reduction
锻件,锻造	forging
半精密锻件	semi-finished forging
长轴锻件	long forging
锻件错移	forging mismatch
锻件平面	forging plane
锻件图	forging drawing
锻件周长	forging circumference
精锻件	finished forging
模锻件	die forging
普通模锻件	conventional die forging
倾斜锻件	tiled forging

115

无斜度精密模锻件	precision no draft forging
圆盘类锻件	forging of round shapes
轴对称锻件	axis-symmetrical forging
飞边	flash
飞边仓部	flash-bin
飞边桥部	flash-land
无飞边	flashless
飞刺,毛刺	fin
分模面	parting plane
分模线	parting line
腹板	web
根部半径	root radius
挤压	extrusion
单位挤压力	extrusion pressure
反挤压	backward extrusion
非稳定挤压	non-steady extrusion
复合挤压	combined extrusion
挤压比	extrusion ratio
径向挤压	radial extrusion
静液挤压	hydrostatic extrusion
连续挤压	continuous extrusion
稳定挤压	steady extrusion
型腔冷挤压	cold extrusion of die cavity
正挤压	forward extrusion

加强肋	stiffening rib
角，角度	angle
超前角	angle of advance
倾斜角	angle of obliquity
咬入角	angle of nip
冷镦	cold heading
连皮	slug
拱形连皮	arched slug
平底连皮	flat-bottomed slug
斜形连皮	sloping slug
毛坯图	block diagram
模具过充满	die over-filling
内圆角	filleted corner
内圆角半径	fillet radius
钳料	tong hold
切边线	trim line
双击镦锻	double-blow heading
投影面积	projected area
图	drawing
草图	rough drawing
锻件图	forging drawing
放大图	enlarged drawing
截面图	cross sectional drawing
冷锻件图	cold forging drawing

热锻件图	hot forging drawing
缩尺图	scale drawing
直径图	diameter drawing
外形,轮廓线	contour
外圆角半径	corner radius
未充满	underfill
斜度	draft
反向斜度,倒锥度	back draft
阶梯模锻斜度	stepped die draft
模锻斜度	die draft
内模锻斜度	inside die draft
外模锻斜度	outside die draft
选配斜度	matching draft
压坯,压块	pre-form
粉末冶金毛坯	powder-metal pre-form
铸造毛坯	cast pre-form
余面	surplus
质量分布曲线	mass distribution curve
中性表面	neutral surface
重复联合镦锻	combined forging and heading
周边,周线	periphery
周边长度	peripheral length
主轴线	principal axis line

13.2 冲压工艺　　Stamping Technology

凹口	notch
变薄拉深	ironing
成形极限图	forming limit diagram
冲孔	punching/piercing
冲裁,落料	blanking
冲裁间隙	blanking clearance
搭边	scrap bridge/web
搭边宽度	width of web
对向凹模冲裁	blanking with opposed dies
翻边	flanging
翻边系数	flanging coefficient
翻孔	hole flanging/plunging
翻孔系数	hole flanging coefficient
负间隙冲裁	negative clearance blanking
光亮带	burnish zone
辊形	roll forming
回弹	spring back
剪裂带	fracture/torn surface
矫直	straightening
精密冲压	fine stamping
精密冲裁	fine blanking
聚氨脂冲裁	polyrethane pad blanking
卷圆,卷边	curling

扩口	flaring/expanding
扩口系数	expanding coefficient
拉深	drawing
充液拉深	hydro-mechanical drawing
多次拉深	multi-stage drawing
反拉深	back/reverse drawing
局部加热拉深	locally-heated drawing
局部冷却拉深	locally-cooled drawing
拉深次数	drawing number
拉深系数	drawing coefficient
橡皮拉深	rubber drawing
液压拉深	hydraulic drawing
预拉深	pre-drawing
正拉深	forward drawing
拉深筋	draw/break bead
拉形	stretch forming
料距	pitch
毛刺	burr
内皱	internal wrinkles
扭曲	twisting
排样,排列	arrangement/layout
单行排列	single arrangement/layout
多行排列	multiple arrangement/layout
混合排列	miscellaneous arrangement/layout

交错排列	zigzag arrangement/layout
坯料排样	blank arrangement/layout
双行排列	double arrangement/layout
斜排	angular arrangement/layout
直对排	straight reverse arrangement/layout
直排	straight arrangement/layout
剖切	parting
起伏	embossing
起皱	wrinkles/wrinkling
切断	cut-off/shearing
切口	lancing
送料节距	feeding pitch
缩口	necking
缩口系数	necking coefficient
塌角	rollover/shear droop
凸耳	earing
外皱	external wrinkles
弯曲	bending
纯塑性弯曲	pure plastic bending
纯弯曲	pure bending
弹-塑性弯曲	elasto-plastic bending
反向弯曲	reversed bending
辊弯	roll bending

拉弯	stretch bending
立体纯塑性弯曲	three-dimensional pure plastic bending
平面弯曲	uniplanar bending
线性纯塑性弯曲	linear pure plastic bending
校整弯曲	bending with sizing
弯曲件展开长度	blank length of bends
弯曲角	bend angle
弯曲线	bend line
危险断面	critical section
相对高度	relative height
相对厚度	relative thickness
相对弯曲半径	relative bending radius
相对转角半径	relative radius
橡皮冲裁	rubber pad blanking
小间隙圆角刃口冲裁	small clearance-round edge blanking
校平	flattening
修边余量	trimming allowance
压边力	blank-holding force
压印	coining
胀形	bulging
胀形系数	bulge coefficient
折边	folding
整形	sizing

整修	shaving
中性层	neutral line
最小弯曲半径	minimum bending radius

14 锻压模具　　Forging-Stamping Dies

14.1　锻造模具　　Forging Dies

仓部	warehouse
承击面	cushion faces
冲子	punch
大锤	sledge hammes
垫模	cushion type die
锻模	forging die
锻模中心	center of the forging die
镦粗垫板	upset block
镦粗台	upset die
剁刀	triangular chisel
飞边槽	groove for flash
分模面	parting area/die parting
分模线	parting line/die parting line
合模	die assembly
夹钳	tongs
检验角	match edge
键槽中心线	central line of the key slot

扣模	buckling die
马架	mandrel supporter
模壁厚度	wall thickness between impression
模膛排列	arrangement of die cavity
模膛深度	depth of die cavity
模膛中心	center of die cavity
劈料台	cleaver divider
钳口	gate
桥部	bridge
切断模膛	cutter
手锤	hand hammer
摔子	tups/swager
锁扣	lock
胎模	loose tooling
套模	cover die
铁砧	anvil
镶块锻模	inserted forging die
芯轴	mandrel
压扁台	flatter
燕尾	shank/dovetail
燕尾中心线	central line of dovetail
预锻模膛	blocker/rougher/blocking
中间模座	intermediate die socket/die holder

14.2 冲压模具　　Stamping Dies

凹模	die/female die
凹模固定板	die block
侧压板	side guide plate
衬套,导套	bushing
齿圈	serrated ring
弹压导板	spring guide plate
挡料销	stop pin/locating pin
导料板	stock guide plate
导向板	guide plate
导正销	pilot pin
导柱	pillar/guide pin
垫板	bolster plate
顶杆	roof bar
顶件器	ejector
顶销	ejector pin
定距侧刀	pilot punch
定位板	locating plate
定位销	locating pin
浮动式模柄	self-centering shank
滚珠导套	ball-bearing bushing
滚珠导柱	ball-bearing type die set guidepin
可卸式导套	demountable guild bushing
可卸式导柱	demountable guide

14 锻压模具

模柄	stalk/shank
模架	die sets/subpress
模具	die
成形模	forming die
冲孔模	punching/piercing die
导板模	guide plate die
导柱模	guide post type die
低熔合金模	low-melting point alloys die
翻边模	flanging die
翻孔模	plunging/hole flanging die
分离模	cutting die
复合模	compound die
钢带模	steel strip die
夹板模	template/steel plate die
简单模	plane die
矫直模	straightening die
卷圆(圈)模	edge rolling die
扩口模	expanding/flaring die
拉深模	drawing die
连续模	progressive die
落料模	blanking die
扭曲模	twisting die
剖切模	sectioning die
切断模	cut-off/shearing die

缩口模	necking die
弯曲模	bending die
无导向模	guide-less die
橡皮模	rubber die
校平模	flattening die
锌基合金模	zinc alloy die
旋压模	spinning die
压印模	coining die
胀形模	bulging die
折边模	folding die
整修模	shaving die
专用模	special purpose die
自动模	transfer die
模具工作部分	working portion of die
模具寿命	die life
模具最小闭合高度	minimum shut height
柔性模	flexible die
上模板	upper bolster
通用模架	universal die sets
凸（凹）模圆角半径	punch (die) radius
凸凹模	punch-die
凸模	punch/male die
凸模固定板	punch plate
推杆	knock-out pin

推件器	knock out/knock-out
下模板	lower bolster
镶块式模	sectional die
斜刃凸模	bevelled punch
卸料板	stripper plate
压边圈	drawing ring/blank holder
压力中心	center of die load
整体式模	solid die
组合式模	combined die

14.3 其他模具　　Other Dies

爆炸成形模	explode forming die
超塑性成形模	super-plastic forming die
成形冲头	forming punch
成形角	forming angle
初镦冲头	preform heading punch
导板	guide plate
电磁成形模	electromagnetic forming die
定径带	calibrating straight
镦锻模具	heading-forging die
镦制凹模	heading die
镦制冲头	heading punch
反挤压	backward extrusion
反压滚轮	counter-pressure roller

分式凹模	combined heading die
复合挤压	combined extrusion
撬棒	pry bar
辊锻模	roll forming die
辊锻线	line of roll forming
弧形拉模	curvilincal drawing die
基本导程	basic lead
基本螺距	basic screw-pitch
挤压模具	extrusion die
开式凹模	open heading die
靠模板	template
螺旋槽	helical groove
平丝板	screw die
双锥面旋轮	double tapered roller
台阶旋轮	stepped roller
凸(凹)模工作带	working straight of punch/die
楔横轧模	wedge rolling tool
楔展角	spreading angle
旋轮	spinning roller
旋轮圆角半径	roller working radius/roller nose radius
旋压芯模	spinning mandrel
预应力圈(缩紧环)	shrink ring
圆丝板	round die
轧辊	roller

整体式凹模	solid die
正挤压	forward extrusion
锥形拉模	conical drawing die
组合式凹模	combination die
最佳挤压凹模角	optimum extrusion angle of die

15 锻压辅助装置 Forging-Stamping Auxillaries

15.1 摩擦与润滑　Friction and Lubrication

半液体润滑	semi-liquid lubrication
边界润滑	boundary lubrication
玻璃润滑剂	glass-lubricant
草酸盐处理	oxalating/oxalate treatment
干燥膜润滑剂	dry film lubricant
固体润滑	solid lubrication
合成润滑剂	synthetic lubricants
极压添加剂	extreme-pressure additive
临界润滑膜厚度	critical thickness of oil film
磷酸盐处理	phosphating/phspohate treatment
摩擦	friction
动摩擦	dynamic friction
固体摩擦	solid friction
静摩擦	statical friction
摩擦功率	friction power
摩擦力	friction force

摩擦热	friction heat
摩擦系数	coefficient of friction
内摩擦	internal friction
外摩擦	external friction
液体摩擦	fluid friction
磨损	wear
磨损率	wear rate
强制润滑	force-feed lubrication
润滑	lubrication
润滑剂	lubricant
润滑膜	lubricant film
润滑性	lubricity
水基石墨润滑剂	colloidal graphite mixed with water
水溶性润滑剂	soluble lubricant
添加剂	additive
液体动力润滑	hydrodynamic lubrication
液体静力润滑	hydrostatic lubrication
液体润滑	liquid lubrication
油基润滑剂	oil lubricant
皂化处理	soap/emulsion treatment

15.2 清理装置　　Cleaning Device

电弧气割清理	arc air gouging clearage
风铲清理	pneumatic chipping

高压喷洗	high pressure hydro-peening
滚筒清理	roller cleaning
抛丸清理	throw shot cleaning
喷砂清理	sand blast cleaning
喷丸清理	shot blast cleaning
砂轮清理	clearage with grinding
酸洗	dipping/pickling
油冷机	oil cold machine
震动清理	vibratory cleaning

15.3 机械化与自动化　Mechanization and Automation

搬运机器人	transport robot
板式输送机	plate conveyer
冲压自动线	stamping automated line
出件装置	output device
电气联动式自动化装置	electrical interlocking automatic device
电液转换器	electro-hydraulic transducer
调头转台	turn-around table/turn table
锻造操作机	manipulator for forging
锻造翻钢机	forging manipulator
锻造自动线	automated forging line
翻转装置	turn-over device

废料处理装置	waste treatment device
钩式送料装置	hook feeder device
辊式送料装置	roller feeder device
滚道	roller conveyer
滑道	slipway
回转式送料装置	rotary feeder device
机械联动式自动化装置	mechanical interlocking automatic device
机械手	robot/mechanical hand
夹板式送料装置	holding plate feeder device
夹持式送料装置	gripper feeder device
接件装置	workpiece catcher device
开卷装置	unwinder
控制器	controller
快锻操作机	quick forging manipulator
快速换模装置	quick die change device
快速换砧装置	quick anvil change device
理件装置	workpiece arrangement device
链式输送机	chain conveter
螺旋振动上料器	helical vibration feeder
皮带输送机	belt conveyer
三坐标夹板式送料装置	three dimension holding plate feeder device
上料机构	feed mechanism

送料节距	feeding pitch
推料机构	pusher
位移脉冲转换器	displacement pulse transverter
蜗杆凸轮辊式送料装置	worm-cam roller feeder device
校平装置	straightener
悬挂输送链	suspension conveyer
旋转镦粗台	rotating upset table
闸门式送料装置	gate feeder device
真空吸料装置	vacuum absorbing device
自动保护装置	automatic safety device
自动可锤装置	device of automatic operating hammer

16 锻压件缺陷 Defect of Forging and Stamping Parts

16.1 锻件缺陷　　Forging Defect

凹痕	dent
凹陷	sunken
白点	flake/lemon spot
超声探伤	supersonic inspection
充填不满	under-filling
锤痕	hammer trace
断裂	fracture
层状断口	lamellar fracture
脆性断裂	brittle fracture
解理断裂	cleavage fracture
萘状断口	naphthalene fracture
石状断口	rock-candy fracture
微孔集聚型断口	microvoid coalescence fracture
延性断裂	tough/gliding/ductile fracture
光谱仪	spectrograph
龟裂	crazing

夹层	interlayer
溅疤	splash scar
角裂	chink
结疤	scab
亮线	bright line
裂纹	crack
发纹	hairline crack
内裂	internal crack
中心横向裂	central and longitudinal crack
中心裂纹	chevron crack
硫印	sulphur
模锻不足	under-pressing
偏析	segregation
气泡	blow hole
缺肉	under-fill
疏松	porosity
缩孔	shrinkage cavity
塌陷	subsidence
探伤仪	flaw detector
凸起	heave
荧光探伤	fluorescent inspection
折叠	overlap/lap

16.2 冲压件缺陷　　Stamping Defect

表面擦伤	surface scratch
波纹	ripple
不直	out-of-straight
侧壁凹陷	sunken of side wall
侧壁裂纹	crack of side wall
侧壁纵向划痕	longitudinal scratch of side wall
尺寸偏差	dimension deviation
单边起皱	unilateral wrinkle
底部凹进	indentation at the bottom
底部不平	uneven bottom
底面翘曲	deflection/warping of bottom
断裂面不直	non-straight of breakage face
拱弯	arch bent
光亮带	burnish zone
过渡圆角不平滑	non-smoothing of transitional fillet
回弹	spring back
剪裂带	fracture/torn surface
孔变形	deformation of hole
孔不同心	non-coaxial of hole
口部开裂	crack of oral area
拉深破裂	drawing fracture
裂口	cracking
裂纹	crack

隆起	upheaval
毛刺过大	burr oversize
内圆角鼓气	air-blowing of filleted corner
破裂	fracture
翘起	warp
翘曲	deflection/warping
撕裂	split
塌角过大	corner collapse oversize
筒身内皱	inner wrinkle of tube body
凸出	bulge
凸耳壁厚不均	wall thickness non-average of projecting lug
凸缘起皱	flange wrinkle
歪斜	skew
弯曲端部凸出	bulge of bending end
弯曲角回弹	spring back of bending angle
弯曲线歪斜	skew of bending line
直边侧壁裂纹	jamb crack of straight wall
皱纹	wrinkle
皱褶	fold

17 模具加工及机器 Die Working and Machine

不圆度,椭圆度	out-of-roundness
车床	lathe/turning machine
尺寸	size
标准尺寸	standard size
公称尺寸	intended/specified size
基本尺寸	basic size
极限尺寸	limiting size
最大极限尺寸	maximum limiting size
最小极限尺寸	minimum limiting size
临界尺寸	critical size
名义尺寸	nominal size
全尺寸,原尺寸	full size
实际尺寸	actual/real size
有效尺寸	effective size
正确尺寸	just size
自由尺寸	free size
总尺寸,轮廓尺寸	overall size

最后净尺寸	finished size
垂直度	verticality
粗糙度,不平（整）度	roughness
倒棱,倒角	chamfer
电化学腐蚀	electro-chemical corrosion
电介成形加工	dielectric forming
对称度	degree of symmetry
多轴钻床	multiple（spindle）drill
工具,刀具	tool
成形车刀	formed turning tool
成形刀	forming tool
刮刀	scraping tool
刨刀	planer tool
镗孔车刀	boring tool
弯头车刀	angular tool
铣刀	milling tool
弓钻	strap drill
公差	tolerance
超出公差	out-of-tolerance
尺寸公差	dimensional tolerance
错移公差	offset tolerance
单向公差	unilateral tolerance
孔径公差	hole tolerance
配合公差	fit tolerance

17 模具加工及机器

双方公差	bilateral tolerance
位置公差,安装公差	location tolerance
小公差	close tolerance
直径公差	diameter tolerance
制造公差	manufacturing tolerance
光洁度	degree of finish
光泽（滑）	gloss
机床	machine
插床	vertical planing machine
电火花加工机床	electro discharge machining (EDM)
	electro sparking machine
电解加工机床	electro-chemical machine
仿形铣床	form milling machine
仿型铣床	copy milling machine
划线机	ruling machine
金属加工机床	metal working machine
金属切削机床	metal cutting machine
靠模磨床	profile grinding machine
靠模铣床	profile milling machine
立式车床	merry-go-round machine
六角车床	turret machine
龙门刨床	planing machine
磨床	grinding machine

牛头刨床	shaping machine
台式机床	bench drilling machine
镗床	boring machine
外圆磨床	external grinding machine
万能磨床	universal grinding machine
卧式刨床	horizontal planing machine
卧式铣床	horizontal milling machine
卧式钻床	horizontal boring machine
无心磨床	centerless grinding machine
铣床	milling machine
样机	model machine
摇臂钻床	radial drilling machine
自动机床	automatic machine
钻床	drill machine
基线,基面,基准	base
精度,精确度	accuracy
测量精度	accuracy in measurement
尺寸精度	accuracy of size
尺寸精度	dimensional accuracy
读数精度	accuracy of reading
几何精确度	geometrical accuracy
计算精确度	accuracy in computation
绝对精确度	absolute accuracy
可达精度	obtainable accuracy

实际精度	available accuracy
相对精度	relative accuracy
校准精度	calibration accuracy
仪表精确度	accuracy of instrument
径向	radial direction
卡尺,游标卡尺	callipers
刻磨	cut
立体的	three-dimensional
量规	gauge
六角钻床	turret drill
抛光,擦光	brightening/buffing
配合	fit
粗配合	coarse fit
过渡配合	transition fit
滑动配合	easy-push fit
混合配合	complicated fit
间隙配合	clearance fit
紧滑配合	close working fit
紧配合	close/tight fit
紧张配合	close running fit
静配合	stationary fit
冷缩配合	contraction fit
膨胀配合	expansion fit
强力配合	forced fit

松配合	easy fit
松转配合	coarse clearance fit
推入配合	drive fit
圆锥连接配合	conical fit
自由配合	free fit
偏差,误差	deviation
标准偏差	standard deviation
基本偏差	basic deviation
均方偏差	mean-square deviation
平均偏差	mean deviation
偏差,余量,公差	allowance
尺寸上偏差	allowance above nominal size
尺寸下偏差	allowance below nominal size
调整余量,装配公差	fitting allowance
负公差	negative allowance
加工余量	machining allowance
精加工余量	allowance for finish
磨损余量	wear allowance
磨削余量	grinding allowance
切削余量	chipping allowance
上偏差	over allowance
无余量,无公差	zero allowance
下偏差	under allowance
正公差	positive allowance

重磨余量	regrinding allowance
平面	plane
基础平面	basic plane
剖面	cut/section plane
水平面	horizontal plane
斜平面	tapered plane
中心平面	central plane
平面度,平整度	flatness/planeness
平行度	parallelism
千分尺	micrometer
切向	tangential direction
全部	finish all over
同心度	concentricity
线,直线	line
尺寸线	dimension line
垂直线	vertical line
弧线	arc line
轮廓线	contour line
曲线	curved line
实线	actual/full/real line
水平线	horizontal line
同轴线	coaxial line
斜线	skew line
虚线,点线	dash/dotted/hidden/imaginary

	line
直线	right/straight line
中心线	center/central line
斜度,斜面	incline
样板	template
圆柱度	cylindricity
轴向	axial direction
钻,钻头	drill
空心钻	hollow drill
扩孔钻	expanding drill
平钻	chucking/flat drill
球头钻	rose drill
深孔钻	depth/deep-hole drill
中心钻	centering drill

18 金属学及热处理　Metallography and Heat Treatment

奥氏体	austenite
奥氏体回火,等温淬火	austempering
奥氏体形变淬火	ausform hardening
包晶的	peritectic
边界	boundary
残余奥氏体	retained austenite
层状的	flaky
超晶组织	superstructure
成分	composition
成核	conception
初结晶	primary crystal
处理,加工	treatment
低温处理	subzero treatment
高频处理	high-frequency treatment
高温热处理	high heat treatment
固溶热处理	solid solution treatment
化学热处理	chemical heat treatment

激光热处理	laser heat treatment
形变热处理	strain heat treatment
粗晶粒的	coarse-grained
脆性	brittleness
红脆	red brittleness
蓝脆	blue brittleness
冷脆	cold brittleness
热脆	hot brittleness
淬火	quenching
淬火剂	quenching medium
淬火浴	quenching bath
淬硬试验	jominy
淬硬性	harden-ability
单晶体	single crystal
氮化法	nitriding
点	point
断裂点	breakaway point
共晶点,低熔点	eutectic point
共析点	eutectoid point
晶格点缺陷	defect of lattice point
临界点	critical point
凝固点	freezing point
电子化合物	electronic compound
调质	improvement

断口	fracture
锻后处理	post-forging treatment
锻后工序	post-forging operation
多晶体的	polycrystalline
多晶型	polymorphism
二硫化钼	molybdenum disulphide
二元的	binary
发蓝	bluing
方向性	anisotropy
非晶质的	amorphous
分解	dissociation
分离,偏析	segregation
格晶	lattice
密排六方晶格	close-packed cubic lattice
面心立方晶格	face centered cubic lattice
体心立方晶格	body centered cubic lattice
共价键	covalent bond
共晶体	eutectic
共析体	eutectoid
固溶体	solid solution
固态线	solidus
固体	solid
过饱和	supersaturation
过共晶的	hypreutectic

核心	core
宏观组织	macro-structure
后处理	post treatment
滑移,滑动	dragging
化合物	chemical compound
回复	recovery
回火	tempering
混合物	mixture
畸变	distortion
价	valence
结晶的	crystalline
结晶面	crystallographic plane
介质(剂)	medium
金相学	metallography
金属键	metallic bond
晶格,晶胞	mesh
晶核	nucleus
晶界	crystal bondary
晶粒	grain
晶粒长大	grain growth
晶粒度,晶粒大小	grain size
本质晶粒度	inherent grain size
实际晶粒度	actual grain size
晶粒形成	grain formation

晶粒直径	grain diameter
空穴	cavity
莱氏体	lidiburite
冷却曲线	cooling curve
离子的	ionic
离子键	ionic bond
连续冷却	continuous cooling
临界点	critical point
临界冷却速度	critical cooling rate
临界体积	critical volume
临界温度	critical temperature
临界压力	critical pressure
临界状态	critical state
六面的	hexagonal
马氏体	martensite
磨光	polish
凝固	solidification
浓度	concentration
片状的	platy
平衡	equilibrium
前处理	preliminary treatment
倾向	tendency
氰化法	cyaniding
球化	spheroidizing

屈氏体	troosite
回火屈氏体	temper troosite
针状屈氏体	acicular troosite
热处理	heat treatment
韧性	toughness
冲击韧性	impact toughness
切口韧性	notch toughness
溶解,溶液	solution
蠕变	creep
渗铬	chromising
渗碳	carburization
表面渗碳	surface carburization
固体渗碳	solid carburization
局部渗碳	local carburization
气体渗碳	gas carburization
渗碳剂	carburizer
渗碳体	cementite
石墨	graphite
石墨化	graphitization
时效	age/aging
时效处理	age/aging treatment
时效裂纹	age/aging crack
时效时间	age/aging time
时效效应	age/aging effect

蚀刻	etch
酸蚀刻	acid etch
试样	sample
索氏体	sorbite
碳化物	carbide
特性,性能,性质	property
铁素体	ferite
同素异构性	allotropy
涂料,涂层	coating
退火	annealing
等温退火	isothermal annealing
球化退火	spheroidizing annealing
完全退火	full annealing
网状物	grid/net
微观组织	micro-structure
位错	dislocation
析出,沉淀	separating
析出	separating
细晶粒的	fine-grained
纤维	fibre
显微镜	microscope
相变	phase change
相律	phase law
相平衡	phase equilibrium

相图,平衡图,状态图	phase diagram
样品	probe
液态线	liquidus
液体,流体	liquid
硬度	hardness
布氏硬度	Brinell hardness
洛氏硬度	Rockwell hardness
莫氏硬度	Mohs hardness
维氏硬度	Vickers hardness
硬化,淬火	hardening
加工硬化	work hardening
油浴	oil bath
预处理	pretreatment
匀均	homogenesity
再结晶	recrystallization
再结晶软化	recrystallization softening
正火	normalization
珠光体	perlite
组成	component
组织,结构	structure
锻造组织	wrought structure
魏氏组织	Widmannstaten structure
原始组织,先前组织	prior structure

19 企业管理及组织 Management and Organization of Enterprise

班,组,队	gang
办公室,营业所	office
中心站	central office
博士	doctor
部门,科,处	division
仓库	bunker/store horse
产量	output
额定生产	rated output
每人班产量	output per man shift
年产量	annual/yearly output
日产量	output per day
总产量	total output
产品,产量	yield
安全产量	safe yield
计算产额	counting yield
年产量	annual yield
常务理事	standing director

157

超额完成	outperform
车间	shop
动力车间	power shop
锻工车间	forge/hammer shop
辅助车间	auxilary shop
工具车间	tool-making shop
锅炉车间	boiler shop
焊接车间	welting shop
机械加工车间	mechanical shop
金工车间	machine shop
木工场	carpenter shop
木模车间	modelling/patter shop
热处理车间	heat treatment shop
设计试制车间	design development shop
试验车间	experimental shop
试制车间	laboratory shop
修理车间	maintenance shop
中心修理车间	central repair shop
铸工车间	foundry shop
装配车间	fitting shop
车间主任	shop master/workshop director
成本,价格,费用	cost
安装费	mounting cost
标准价格	standard cost

成本	final cost
成本单价	cost unit price
工料费	flat cost
基本费用	capital cost
进货价格	prime cost
模具费	die cost
生产成本	production cost
维护费	running cost
原价	first cost
运输费	haulage cost
制造费	fabricating cost
成品,产品	product
半成品	half-finished/semi-finished product
成品	end/finished product
纯产品	straight product
二次产品	after product
副产品	border-line product
特殊产品	specialty product
主要产品	prime product
冲压工	hubber
处,科,股,组	section
财务科	financial section
动力科	power section

工具科	tool-making section
工艺科	technological section
计划科	plan section
技术科	technical section
检验科	control section
设计科	design section
生产科	production section
措施,方案	arrangement
大修理	capital repair
大学	university
调度站	shunting station
定额	quota
锻工	blacksmith
副博士	vice-doctor
副教授	vice-professor
副理事长	vice president
高等专科学校	academy
高级参谋	senior staff officer
高级顾问	high-ranking adviser
革新	innovation
革新者	innovator
工厂设计	factory design
工厂委员会	factory committee
工程师	engineer

电气工程师	electrical engineer
高级工程师	senior engineer
工艺工程师	process engineer
机械工程师	machine engineer
设计工程师	design/project engineer
冶金工程师	metallurgical engineer
主任工程师	chief engineer
助理工程师	assistant engineer
装配工程师	installation engineer
总工程师	general engineer
工业	industry
大规模工业	large-scale industry
轻工业	light industry
重工业	heavy industry
工艺师	technologist
工资	wage
工资等级	wage scale
工资冻结	wage freeze
工资基金	wage fund
名义工资	nominal wage
工作,职务	job
计件工作	piece-work job
兼职工作	part-time job
公司	company

公文,文件	document
出口单据	export document
发货凭单	shipping document
供应	supply
规模,比例尺	scale
锅炉工	kettleman
国际市场价格	world market price
合并	merge
合理化	rationalization
合理化建议	rationalization proposal
合同	contract
合营	jointly owned
合作,协作	cooperation
机器制造厂	machine building works
机械师	mechanist
积累	accumulate
计划,平面图	plan
测量示意图	instrumentation plan
发展计划,开拓计划	development plan
施工平面图	construction plan
索引图	key plan
远景计划,长远计划	far-seeing plan
总计划	general plan
技术员,技师	technician

技术员	operator
讲师	lecturer
交货,交付	delivery
教师	teacher
教授	professor
经理	manager
副经理	assistant manager
总经理	general manager
局,司,处	bureau
局长,处长,所长,董事,厂长,主任	director
代理厂长	acting director
副厂长	vice director
科学院	academy of science
理事长	president
理事会,董事会	directorate
利润	profits
纯利润,净利润	net profits
总利润	gross profits
领班,工长	foreman
模锻工	smith
企业	business
人员	staff
设计师	designer

生产	production
超额生产	excess production
成批生产	batch production
大量生产	mass production
大批生产	high-run/large-lot/large-scale production
单件生产	piece/job-lot production
个体生产	individual production
连续生产	line production
流水线生产	flow-line production
生产成本	production cost
生产计划	production plan
生产力	production forces
小量生产	short-run/small volume production
小批生产	small lot/small serial production
自动化生产	automation production
总产量	cumulative production
生产率	productivity
实习,实践,操作	practice
安全技术,安全规章	safe practice
操作规程	routine practice
业务实习	office practice
实验室	laboratory

收入	receipts
收入和支出	receipts and expenditures
硕士生	postgraduate
司炉工	fireman
统计数据	statistical date
图表,卡片	chart
工艺卡片	process chart
进度表	progress chart
下料工	blanking worker
协会	association/society
协议	agreement
学会	institute
学院	college
研究所	research institution
研究员	research fellow
原始资料	initial date
院士	academician
载荷	load
冲击载荷	impact load
极限载荷	ultimate load
剪切载荷	shear load
静力载荷	static load
扭转载荷	torsional load
偏心载荷	eccentric load

设计载荷	design load
账目,报表	account
折旧率	depreciation rate
支票,检查	check
指标,指数	index
制图室	drawing office
制图板	drawing plate
制图仪器	drawing instrument
制图纸	drawing paper
制图员	draftsman
制造,生产	manufacture
大量生产	quantity manufacture
大批生产	extensive/large scale/wholesale manufacture
小批生产	small lot manufacture
助教,助理	assistant
专家	specialist
资产,财产	assets
固定资产	fixed assets
流动资产	liquid assets
总布置	general layout
组织,编制,机构,协会	organization
设计机构	project organization
文件编制	file organization

19 企业管理及组织

3D 打印	three dimensional printing (TDP)
边界元法	boundary element method (BEM)
并行工程	concurrent engineering (CE)
并行设计	concurrent design (CD)
差错恢复程序	error recovery program (ERP)
产品供应链管理	product supply chain management (PSCM)
产品数据管理	product data management (PDM)
产品信息管理	product information management (PIM)
叠层实体造型	laminated object manufacturing (LOM)
反向工程	reverse engineering (RE)
高级计划管理	advanced plan management (APM)
工程数据管理	engineering data management (EDM)
工程研究报告	engineering research report (ERR)
管理信息系统	management information system (MIS)
基于规则的推理	rule-based reasoning (RBR)
基于模型的推理	model-based reasoning (MBR)
基于人工神经网络技术	based on artifical neutral network technology (BANNT)
基于事例的推理	case-based reasoning (CBR)

基于知识的工程	knowledge-based engineering (KBE)
计算机辅助测试	computer aided test (CAT)
计算机辅助工厂	computer aided factory (CAF)
计算机辅助工程	computer aided engineering (CAE)
计算机辅助工艺规划	computer aided process planning (CAPP)
计算机辅助讲授	computer aided instruction (CAI)
计算机辅助教学	computer aided teaching (CAT)
计算机辅助设计	computer aided design (CAD)
计算机辅助制造	computer aided manufacturing (CAM)
计算机辅助质量	computer aided quality (CAQ)
计算机集成制造系统	computer integrated manufacturing system (CIMS)
计算机数字管理	computer numerical control (CNC)
计算机数字控制	computer numerical control (CNC)
精益生产	lean production (LP)
客户关系管理系统	clientele relation management system (CRMS)
快速原型制造	rapid prototyping maufacturing (RPM)
立体光化造型	stereo lithography apparatus (SLA)
绿色高端制造	green and high-grade manufacturing

	(GHGM)
敏捷制造	agile manufacturing (AM)
模具智能制造	mould intelligent manufacturing (MIM)
排程管理	schedule management (SM)
全面质量管理	total quality management (TQM)
人工智能	artificial intelligence (AI)
人工智能设计	artificial intelligence design (AID)
熔融堆积造型	fused deposition modeling (FDM)
柔性制造系统	flexible manufacturing system (FMS)
上限单元技术	upper bound element technique (UBET)
设计面向测试	design for test (DFT)
设计面向成本	design for cost (DFC)
设计面向制造	design for manufacturing (DFM)
设计面向质量	design for quality (DFQ)
设计面向装配	design for assembly (DFA)
特征技术	feature techonology (FT)
同步工程	simultaneous engineering (SE)
网络系统	network system (NS)
先进制造技术	advanced manufacturing technology (AMT)
协同设计	collaborative design (CD)

虚拟路径	virtual route (VR)
虚拟制造	virtual manufacturing (VM)
选择性激光烧结	selected laser sintering (SLS)
有限单元法	finite element method (FEM)
制造执行系统	manufacturing execute system (MES)
制造自动化系统	manufacturing automated system (MAS)
智能管理	intelligent management (IM)
智能机器人	intelligent robot (IR)
智能设计	intelligent design (ID)
智能制造	intelligent manufacturing (IM)
智能制造系统	intelligent manufacturing system (IMS)
专家系统	expert system (ES)

附 录

附录1 常用化学元素名称表

原子序数	元素名称	元素符号	英　文
1	氢	H	hydrogen
2	氦	He	helium
3	锂	Li	lithium
4	铍	Be	beryllium
5	硼	B	boron
6	碳	C	carbon
7	氮	N	nitrogen
8	氧	O	oxygen
9	氟	F	fluorine
10	氖	Ne	neon
11	钠	Na	sodium
12	镁	Mg	magnesium
13	铝	Al	aluminum
14	硅	Si	silicon
15	磷	P	phosphorus
16	硫	S	sulfur

续 表

原子序数	元素名称	元素符号	英文
17	氯	Cl	chlorine
18	氩	Ar	argon
19	钾	K	potassium
20	钙	Ca	calcium
22	钛	Ti	titanium
23	钒	V	vanadium
24	铬	Cr	chromium
25	锰	Mn	manganese
26	铁	Fe	iron
27	钴	Co	cobalt
28	镍	Ni	nickel
29	铜	Cu	copper
30	锌	Zn	zinc
32	锗	Ge	germanium
33	砷	As	arsenic
34	硒	Se	selenium
35	溴	Br	bromine
40	锆	Zr	zirconium
41	铌	Nb	niobium
42	钼	Mo	molybdenum
45	铑	Rh	Rhodium

续 表

原子序数	元素名称	元素符号	英　文
46	钯	Pd	palladium
47	银	Ag	silver
48	镉	Cd	cadmium
50	锡	Sn	tin
51	锑	Sb	antimony
53	碘	I	iodine
56	钡	Ba	barium
58	铈	Ce	cerium
60	钕	Nd	neodymium
74	钨	W	tungsten
75	铼	Re	rhenium
78	铂	Pt	platinum
79	金	Au	gold
80	汞	Hg	mercury
82	铅	Pb	lead
88	镭	Ra	radium
92	铀	U	uranium

附录2 常用计量单位公制与英制换算表

附表 2-1 长度 length

米(m)	千米(km)	市尺	英寸(in)	英尺(ft)	码(yd)	英里(mi)
1	0.001	3	39.37	3.280 84	1.093 61	0.000 62
1 000	1	3 000	39 370	3 280.84	1 093.61	0.621 37

附表 2-2 面积 area

平方米 (m²)	公亩 (a)	公顷 (ha)	平方千米 (km²)	市亩	英亩 (acre)
1	0.01	0.000 1	0.000 001	0.001 5	0.000 25
100	1	0.01	0.000 1	0.15	0.024 71
10 000	100	1	0.01	15	2.471 06
1 000 000	10 000	100	1	1 500	247.106
4 046.86	40.468 6	0.404 69	0.004 05	6.070 29	1

附表 2-3 体积 volume

英文名称和符号	中文名称	等 量	折合公制
cubic yard (yd³)	立方码	27 立方英尺	0.764 554 m³
cubic foot (ft³)	立方英尺	1 728 立方英寸	28 317 cm³
cubic inch (in³)	立方英寸	1/1 728 立方英尺	16.387 1 cm³

附表 2-4　容积[液体]volume [liquid]

升(L)	英液盎司(fl oz)	美液盎司(fl oz)	英加仑(gal)	美加仑(gal)
1	35.198 8	34.164 8	0.219 97	0.264 17

附表 2-5　质量 mass

千克(kg)	吨(t)	市　斤	盎司(oz)	磅(lb)
1	0.001	2	35.274 0	2.204 62
1 000	1	2 000	35 274.0	2 204.62

1 磅 = 0.453 592 45 千克

附录3 锻压设备技术参数举例

1. **空气锤主要技术参数** Main technical parameter of air hammer

落下部分重量(kg)　　Falling parts weight
打击能量(不小于)(J)　　Blowing energy (not less than)
锤头每分钟打击次数(次/min)　　Ram blow per minute
工作区间高度(mm)　　Working zone height
锤杆中心线至锤身距离(mm)　　Distance from center line of piston rod to ram body
上,下砧块平面尺寸(mm)　　Top and bottom anvil plane dimensions
砧座重量(kg)　　Anvil block weight

2. **蒸汽-空气自由锻锤主要技术参数** Main technical parameter of steam-air smith hammer

落下部分重量(t)　　Falling parts weight
最大打击能量(不小于)(kJ)　　Max blowing energy (not less than)

锤头每分钟打击次数(次/min)	Ram Blows per minute
锤头最大行程(mm)	Ram max stroke
气缸直径(mm)	Cylinder diameter
锤杆直径(mm)	Piston rod diameter
下砧至地面距离(mm)	Distance from top surface of lower anvil to floor
两立柱间距离(mm)	Distance between uprights
上砧面尺寸(mm)	Upper anvil surface dimension
下砧面尺寸(mm)	Lower anvil surface dimension
导轨间距离(mm)	Distance between guides
蒸汽消耗量(t/h)	Steam consumption
砧座重量(t)	Anvil block weight
机器总重(t)	Machine total weight
外形尺寸(长×宽×地面上高)(mm)	Overall dimensions (length×width×height above floor)

3. 蒸汽-空气模锻锤主要技术参数

Main technical parameter of steam-air die forging hammer

落下部分重量(t)	Falling parts weight
最大打击能量(kJ)	Max blowing energy (not less than)
锤头最大行程(mm)	Ram max stroke
锻模最小闭合高度(mm)	Forging die min daylight height

导轨间距(mm)	Distance between guides
锤头前后方向长度(mm)	Length between front and back of ram
模座前后方向长度(mm)	Length between front and back of die base
锤头每分钟打击次数(次/min)	Ram blows per minute
蒸汽绝对压力(10^3 Pa)	Steam absolute pressure
蒸汽允许温度(℃)	Steam allow temperature
砧座重量(t)	Anvil block weight
总重量(不带砧座)(t)	Total weight (without anvil block)
外形尺寸(前后×左右×地面上高)(mm)	Overall dimension (depth×width×height above floor)

4. 双盘摩擦螺旋压力机主要技术参数 Main technical parameter of double disc friction press

型号	Model
公称压力(kN)	Nominal pressure
打击能量(kJ)	Blowing energy
滑块行程(mm)	Slide stroke
行程次数(次/min)	Stroke number
封闭高度(mm)	Shut height
垫板厚度(mm)	Setting thickness

工作台尺寸(mm)	Table surface dimension
导轨间距(mm)	Distance between guides
电动机功率(kW)	Motor capacity
外形尺寸(前后×左右×地面上高)(mm)	Overall dimension (depth × width × height above floor)
机器总重量(t)	Machine total weight

5. 热模锻压力机主要技术参数　　Main technical parameter of hot die forging press

型号	Model
公称压力(kN)	Nominal pressure
滑块行程(mm)	Slide stroke
滑块行程次数(次/min)	Slide stroke number
封闭高度(mm)	Shut height
工作台尺寸:左右(mm) 前后(mm)	Table size, width depth
滑块尺寸:左右(mm) 前后(mm)	Slide size, width depth
滑块行程位置调节量(mm)	Adjustment of slide stroke

上顶料器：压力(kN)	Top ejector, force
行程(mm)	stroke
下顶料器：压力(kN)	Bottom ejector, force
行程(mm)	stroke
主电机：功率(kW)	Main motor, power
转速(r/min)	speed
机器总重量(t)	Machine total weight

6. 精压机主要技术参数　　Main technical parameter of double disc friction press

型号	Model
公称压力(kN)	Nominal pressure
公称压力行程(mm)	Nominal pressure stroke
滑块行程(mm)	Slide stroke
滑块行程次数(次/min)	Slide stroke number
最大装模高度(mm)	Max die set height
装模高度调节量(mm)	Die set height adjustment

7. 闭式单点压力机主要技术参数　　Main technical parameter of straight side single point press

型号	Model
公称压力(kN)	Nominal pressure
滑块行程(mm)	Slide stroke

滑块行程次数(次/min)	Slide stroke number
最大装模高度(mm)	Max die set height
装模高度调节量(mm)	Die set height adjustment
工作台尺寸：左右(mm)	Table size, width
前后(mm)	depth
滑块底面尺寸(mm)	Slide bottom face dimension
主电机功率(kW)	Main motor power
机器总重量(t)	Machine total weight

8. 双动薄板冲压液压机主要技术参数

Main technical parameter of double action thin-slab stamping hydraulic press

型号	Model
总压力(kN)	Total pressure
液压力(kN)	Hydraulic force
拉伸滑块压力(kN)	Drawing slide pressure
压边滑块压力(kN)	Blank holder slide pressure
压边缸液压力(kN)	Blank holder cylinder pressure
拉伸滑块最大行程(mm)	Drawing slide max stroke
压边滑块最大行程(mm)	Blank holder slide max stroke
最大拉伸深度(mm)	Max drawing depth

工作台尺寸(前后×左右)(mm)	Working table size (FB×LR)
液压垫压力(kN)	Hydraulic cushion pressure
主电机功率(kW)	Main motor power
外形尺寸(前后×左右×地面上高)(mm)	Overall dimensions (depth × width × height above floor)
床机重量(t)	Machine weight

9. 下传动双动拉深压力机主要技术参数 Main technical parameter of under-drive type double action drawing press

型号	Model
拉深滑块公称压力(kN)	Drawing slide nominal pressure
压边滑块公称压力(kN)	Blank holder slide nominal pressure
滑块行程次数(次/min)	Slide strokes number
最大坯料直径(mm)	Max blank diameter
最大拉深深度(mm)	Max drawing depth
最大拉深直径(mm)	Max drawing diameter
拉深滑块行程(mm)	Drawing slide stroke
导轨间距离(mm)	Distance between uprights
工作台行程(mm)	Working table stroke

压边滑块底面至工作台最大距离(mm)（压边滑块及工作台均在最高位置）	Max distance, bottom of holder to working table (all at top dead positions)
工作台尺寸(前后×左右)(mm)	Working table size (FB×LR)
工作台孔径(mm)	Diameter of hole in working table
装模杆螺纹	Thread of adapter for connecting upper die
装模杆螺纹长度(mm)	Thread length of adapter
拉深滑块最大调节量(mm)	Drawing slide max adjustment
压边滑块最大调节量(mm)	Blank holder max adjustment
主电动机功率(kW)	Main motor power
调整电动机功率(kW)	Adjustment motor power
外形尺寸(前后×左右×地面上高)(mm)	Overall dimensions (FB×LR×H)
机器重量(约)(t)	Machine weight, (approx)

10. 曲柄式冷挤压机主要技术参数

Main technical parameter of crank cold extrusion press

型号	Model

公称压力(kN)	Nominal pressure
公称压力行程(mm)	Nominal pressure stroke
滑块行程(mm)	Slide stroke
滑块行程次数(次/min)	Slide stroke number
最大装模高度(mm)	Max die set height
装模高度调节量(mm)	Die set height adjustment
工作台尺寸(前后×左右)(mm)	Working table size (FB×LR)
滑块底面尺寸(前后×左右)(mm)	Slide bottom face size (FB×LR)
下顶出器：行程(mm) 顶出力(kN)	Bottom ejector, stroke ejector force
主电动机功率(kW)	Main motor power
机器重量(t)	Machine weight

11. 板料多工位压力机主要技术参数

Main technical parameter of blank multi-position press

型号	Model
公称压力(kN)	Nominal pressure
公称压力行程(mm)	Nominal pressure stroke
滑块行程(mm)	Slide stroke
滑块行程次数(次/	Slide stroke number

min)
工位数　　　　　　　　Position number
工位间距(mm)　　　　　Distance between positions
最大装模高度(mm)　　　Max die set height

12. 自动多工位冷锻压力机主要技术参数[德国舒勒公司] Main technical parameter of automatic multiposition cold forging press

型号　　　　　　　　　Model
公称压力(kN)　　　　　Nominal pressure
成形工位数　　　　　　Forming position number
线材最大直径(mm)　　　Max diameter of wire rod
凹模模具直径(mm)　　　Die diameter of mould
凸模模具直径(mm)　　　Die diameter of punch
切断长度最大值(mm)　　Max value of cut off length
凹模顶出器行程(mm)　　Mould ejector stroke
凸模顶出器行程(mm)　　Punch ejector stroke
生产率(ppm)　　　　　　Productivity
机器重量(t)　　　　　　Machine weight

13. 双支承辊锻机主要技术参数 Main technical parameter of double bearing roll

型号　　　　　　　　　Model

锻模公称直径(mm)	Forging die nominal diameter
公称压力(kN)	Nominal pressure
锻辊直径(mm)	Forging roll diameter
锻辊可用长度(mm)	Useful length of forging roll
锻辊转速(r/min)	Forging roll speed
锻辊中心距调节量(不小于)(mm)	Inter-nodal adjustment of forging roll (not less than)
可锻毛坯长度(mm)	Forging blank length

参考文献

[1] [苏]A.B.米海叶娃等.科技英语最低限度词汇[M].任德,编译.北京:商务印书馆,1961.

[2] 北京外国语学院英语系《汉英词典》编写组.汉英词典[M].北京:商务印书馆,1978.

[3] 洪慎章.汉英锻压专业分类词汇[M].上海:上海交通大学教材科,1984.

[4] 史竞.工业炉词典[M].北京:机械工业出版社,1996.

[5] 英国牛津大学出版社字典部.牛津-杜登英汉图解词典[M].北京:化学工业出版社,1984.

[6] 《英汉工程技术词汇》编辑组.英汉工程技术辞汇[M].北京:国防工业出版社,1976.

[7] 中国机械工程学会锻压学会.锻压词典[M].北京:机械工业出版社,1989.